CONDOR COMEBACK

Words by SY MONTGOMERY

Photographs by TIANNE STROMBECK

85

Houghton Mifflin Harcourt
Boston   New York

IN MEMORY OF SARA SEWARD

Text copyright © 2020 by Sy Montgomery
Photographs copyright © 2020 by Tianne Strombeck

All photographs copyright © Tianne Strombeck except the following:
Akinshin/Getty Images (feather photo): pp. 1, 13, 20, 33, 40, 57, 62, 71, 76; Kathleen Conti: p. 64;
Mart_M/Getty Images (feather dingbat): pp. 7, 26; Sy Montgomery: p. 31; US Fish and Wildlife Service,
Joseph Brandt: pp. 71, 73 (both); US Fish and Wildlife Service, Los Angeles Zoo and Botanical
Gardens: p. 79 (top left); US Fish and Wildlife Service, Santa Barbara Zoo: pp. 70, 74 (both);
US Fish and Wildlife Service, Santa Barbara Zoo, Cornell Lab of Ornithology: p. 75.

hmhbooks.com

Map artwork by Holly A. Sullivan
Design by Cara Llewellyn
The text type was set in Minion Pro.
The display type was set in Industrial Gothic and Letterpress Text.

Library of Congress Control Number: 2019020308
ISBN: 978-0-544-81653-4

Manufactured in China
SCP 10 9 8 7 6 5 4 3 2 1
4500796089

CONTENTS

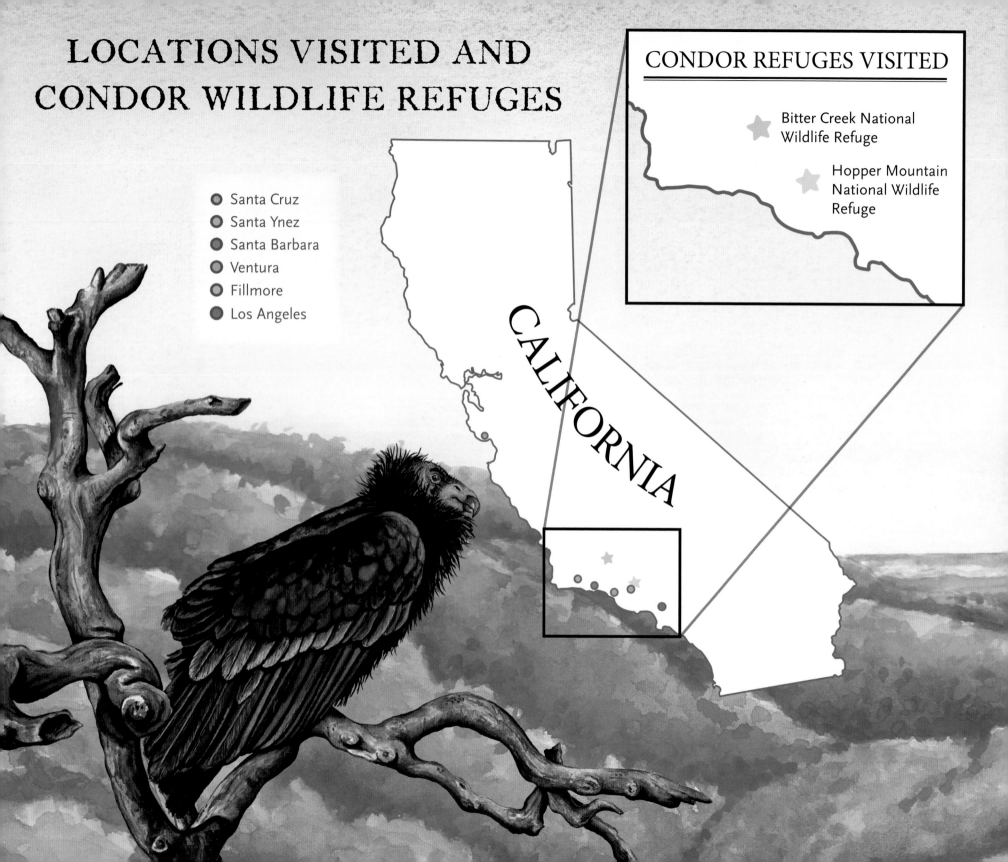

LOCATIONS VISITED AND CONDOR WILDLIFE REFUGES

- Santa Cruz
- Santa Ynez
- Santa Barbara
- Ventura
- Fillmore
- Los Angeles

CALIFORNIA

CONDOR REFUGES VISITED

Bitter Creek National Wildlife Refuge

Hopper Mountain National Wildlife Refuge

CANADA

UNITED STATES

MEXICO

Historic Condor Range

Current Condor Range

HISTORIC VS. CURRENT CONDOR RANGE

Condor 174. The red color on the tag stands for 100 so the other numbers can be larger and easier for biologists to see from a distance.

Chapter One:
AT THE ZOO

SHE NEEDS TO DO NOTHING MORE than stand still to attract a crowd.

Perched on her favorite rock outcropping in the spacious exhibit at the Santa Barbara Zoo, her wings clad in shiny black feathers that rustle like taffeta, California Condor 174 is a giant among birds. She towers at four feet (1.29 meters) tall—taller than the average seven-year-old girl—and weighs nearly thirty pounds (almost fourteen kilograms, or as much as a hundred baseballs). Her species is the largest species of bird in all of North America. Even her feathers are giants: some of them grow two feet (sixty-one centimeters) long. No wonder a group of people—including youngsters smaller than she—has gathered to watch her.

She turns her orange neck and head to face the onlookers. Her red, knowing eyes briefly meet ours. It feels like a meeting of minds. With her stooped posture and bald, wrinkled, jowly head, she looks like a wizened sorceress, a sage, a powerful, wise old woman. When she raises her wings, holding them slightly open, she looks like she's about to give a blessing—or cast a spell.

Then, the magic really happens: she hops twice, flaps thrice, and spreads her wings nearly ten feet (over three meters) wide to sail across her enormous pen.

"Wow! Look how big those wings are!" says a little girl wearing a pink sweatshirt and American flag sneakers.

"Spread your wings!" a bearded dad urges his youngest daughter. Immediately, the little girl and her three siblings rush to compare their arm span to a life-size sign opposite the pen, showing a condor's yawning wingspan.

Thanks to these astonishing wings, a California condor can not only fly at a speed of 55 miles (88 kilometers) an hour but also soar to 15,000 feet (4,572 meters). Even more impressive, a condor can glide for miles without flapping, riding on rising currents of hot air called thermals and steering with just its tail and the tips of its long flight feathers. Condors don't just traverse heaven; they dwell there.

It's easy to see why these birds have thrilled and fascinated people for thousands of years. Once California condors were found in western skies from Canada to Mexico, and some lived as far east as Florida. Native people revered them. To many tribes, the condor was sacred. This was with good reason: Flying so high, the condor sees all. And these birds may live for sixty or more years—long enough to grow wise.

But the California condor was not sacred to Western settlers. Far from it. The newcomers shot the birds for sport. Ranchers accused them—falsely—of killing livestock. By the time conservationists realized condors were disappearing, their slide into extinction seemed unavoidable.

"Aren't they endangered?" a ponytailed woman watching 174 wonders.

"They are *critically* endangered!" answers Dr. Estelle Sandhaus. In fact, Estelle tells the visitor, in 1982 there were fewer than two dozen of them left alive on the planet—and when the last one was captured in Southern California in April 1987, the California condor was officially extinct in the wild.

A firecracker of a woman, standing five feet one inch tall, with shining brown hair, dancing brown eyes, and a laugh as exuberant as a waterfall, 41-year-old Estelle is the Santa Barbara Zoo's director of conservation science. A big part of her job is to help make sure California condors forever grace North American skies.

And that's Condor 174's job, too. Born March 4, 1998 at the San Diego Zoo Safari Park, 174 came to Santa Barbara on October 15, 2012, where she is now serving as a mentor to younger birds. "She's the most dominant bird," Estelle explains. "She's got sass. She's got attitude. She knows she's the boss."

At the moment, 174's mentee is young Condor 603. She's the youngest of the four California condors at the zoo (two others are not on display). Condor 603 was born in the wild but suffered a wing injury and was brought to the zoo. She can fly, but not well. At age three, she's still a child by condor standards. She's got much to learn—including condor manners.

how a young condor should behave around her elders at mealtime.

"They're going to get rabbits today," announces zoo bird keeper Ellie Culip. The condors eat four times a week. (In the wild, they sometimes eat so much they can't fly for several hours, and they might not eat at all for several days afterward.) On today's menu are white rabbits that were obtained from a breeder, humanely killed, then frozen, and thawed.

Ellie walks inside a concrete tunnel built into the artificial rock outcropping in the exhibit. She dons plastic gloves and reaches into a white plastic bucket for the first of the two rabbits. There are two narrow tunnels built into the rock, each just a little longer than Ellie's arm. Ellie will use one of these tunnels to push the food through to the condors on the other side.

Why not just hand the birds the carcass? "We never let them see us with the food," says Ellie. If wild condors are fed by humans, they'll search out people—and that can be dangerous for an entire flock, because they learn from watching each other. And though these condors aren't slated for release—both will probably stay at a zoo for breeding or to mentor other condors—"we don't want to limit their possibilities for the future if management changes," Ellie says.

Condors are social creatures, like people. They like to do things in groups. When some of the captive birds were first moved to an exhibit at the Santa Barbara Zoo, a keeper noted that the whole group, together, carefully plucked every California poppy that was in bloom in the exhibit, and put them all in a pile in the corner. Then the flock moved the pile. The first time one of the zoo's condors landed on the weighing scale, all the other condors then jumped on it. Because togetherness is important, 603's education includes learning

Condors are social birds. Here, 174 teaches her mentee, Condor 603, some table manners.

But it's difficult to fool a condor. An orange face appears at the end of one of the tunnels. It's 174. "They're smart birds," she explains. "They know I'm feeding them. But at least they never see me putting food down!"

As soon as it appears at the other end of the tunnel, 174 instantly grabs the rabbit with her beak. Then Ellie pushes the second rabbit through. This one is for young 603, but, says Ellie, "I would not be surprised if 174 kicks 603 off and wants *her* rabbit, too!"

This is exactly what happens. 174 yanks the second carcass away. She isn't being a bully, she's being a good mentor. Although condor parents lovingly feed their babies, when the chicks get older, they must learn the rules. And one of the most important rules is that, after they leave the nest, youngsters must defer to their elders.

When condor conservationists first started raising captive-bred condor chicks, the adolescents that were released to the wild

were in for a rude awakening. They probably expected to be babied like they were in the zoo. "The hand-raised youngsters weren't tough enough," Estelle explained. "There was no condor culture to teach them respect. They didn't know what to do or quite how to behave like wild condors."

A wild condor will behave toward a youngster just like 174 does toward 603: Raising those formidable wings and showing off her sharp beak, the elder bird chases her

Zookeeper Ellie Culip readies a meal for Condors 603 and 174.

Condor 174 peers into the tunnel, anticipating her meal.

young student away from the carcass again and again. But soon enough, when 174 has had her fill, there will be tasty scraps for 603 to enjoy, left behind on the rock outcropping. There's plenty for both condors.

When Ellie exits the concrete bunker, in an adjacent pen, Veronica, a turkey vulture, hops over to see if the keeper might have a scrap for her. "Is this a baby condor?" one of the bystanders asks.

"No," answers Ellie. "But people ask that a lot. And it's a smart question. Condors are vultures, after all."

And that's part of the uphill battle still being fought to save condors. "Vultures have a stigma," says Estelle. "Some kids are like, 'Ew, *vultures*! They're mean. They're gross. They eat dead things.'" Stories and films often portray vultures as icky. (Though at least Disney's vultures are cool: In the original script for the first *Jungle Book* movie, in 1967, the four vultures were to be voiced by the Beatles! When the Fab Four declined, the scriptwriter changed the singing vultures into a barbershop quartet.)

Estelle and Ellie tell kids that condors, like all vultures, are scavengers. "They're not mean," says Estelle. "They don't even kill!" (Indeed, the author of one magazine article insisted that the condor is "nature's original conscientious objector"—a huge and powerful bird who "could be a killer, but chooses instead to live in peace with his fellow creatures.")

And condors are characters. Some are

Veronica, the turkey vulture.

bold. Some are shy. Some are bossy. Some are brave. When Estelle first met 327, she held the bird during her health check. The next time 327 saw her, the feisty female remembered Estelle as the person who had restrained her—and hissed at her in defiance!

"I love to tell kids how each one has a personality," says Estelle. "They're amazed that a bird can be as cool as a fox or an elephant. But it's true.

"It makes me sad when I hear people say condors are ugly. They need to open their eyes to beauty that's a little bit different," Estelle urges. "I think they're absolutely beautiful!"

Most visitors who spend time with condors agree. One man comes regularly and sets up his easel in front of the condor exhibit to paint them.

Four-year-old Abigail Rose Drennan is just as ardent a fan as he. With her mom, Traci, and often with her brother, eight-year-old James, Abigail has visited the condor pen at the zoo once a week since the day she was born. "They are so big," she says, hopping with excitement, "and I like them!"

Even though she's only four, Abigail knows that the zoo is helping condors. "People have to help animals," she says. "People should care because, if condors are gone, it would be sad."

But there was a time, not long ago, when many people, including some genuine conservationists, believed that condors should never be in a zoo. "The beauty of the California condor lies entirely in its matchless, soaring flight," wrote one of the first researchers to study these birds, Carl Koford.

When he wrote those words, wild condors were already disappearing. But he objected to the idea of capturing them so they could breed in the safety of a zoo. "The California condor in a cage is ugly, pitiful and uninspiring," he claimed. Friends of the Earth founder David Brower thought captive breeding was worse than extinction. Better, he thought, that they should "die with dignity" in the wild.

The naysayers have been proved wrong. Zoos have been key to the condor comeback.

Condor 174 takes a regal stance as she watches zoo visitors.

breed, and raise their young without human help—is one of the most creative, controversial, and concentrated wildlife restoration efforts in conservation history.

TODAY, THE WORLD'S POPULATION of California condors has grown from twenty-two to more than four hundred fifty. A little over a third of them are in zoos like this one, where they are recuperating from illness or injury; teaching younger birds condor culture and etiquette; or mating, laying, or incubating eggs to augment their still small population.

Most of the world's California condors fly free today—not just here in Southern California, but in Central California, Arizona, Utah, and Mexico. But even after half a century of human help, the wild condors still need constant monitoring. People still need to step in to help. Every California condor wears a wing tag with a number so they can be identified and telemetry so they can be located and followed. Every one still needs regular health checks—and sometimes medical treatment—in a battle for survival in a world polluted by human-made garbage and toxic metal.

"People say, 'Fifty years and they're not saved yet?!'" Estelle understands the frus-

"The challenges condors are facing are entirely human in origin," Estelle says, "and the good news is that the answers are also in human hands. We can do it. We know what to do. And this connects the people with the condors, so our visitors can be part of the solution."

The solution is at once simple and complicated. If humans would stop putting just two foreign toxic substances into the environment, the vast majority of premature condor deaths would be prevented. But until people change their ways, folks like Estelle, and many others you'll meet in this book, are working hard to give the magnificent birds a second chance at survival.

The effort to return California condors to the wild—a wild where they can safely live,

tration. At the zoo, Estelle is also working on saving endangered Channel Island foxes. (With three subspecies of the foxes taken off the endangered species list, it's the fastest successful recovery of a federally listed endangered species in American history!)

Why is that project moving forward so much more swiftly than the condor recovery? "Foxes have litters every year. Condors aren't old enough to find a mate till they're six, seven, or older. These are long-lived animals.

This is like the marathon of conservation— while foxes are like the mile run.

"We've hit a lot of major milestones in the program to conserve condors," Estelle notes. "But we're not done yet!"

This book tells the story of the California condors' continuing comeback. Its history contains more plot twists than a mystery novel; the recovery program faced, faltered, and overcame a minefield of obstacles. But there are still more to go. This is the story of

partnerships between birds and people, cooperation between zoos, other nonprofits, and state, federal, and even international governmental organizations. It's a story of roughly equal parts human vision and human blindness. Even when you finish this book, you won't know how the story ends—but you will know some of what it takes to save a species, and what you can do to help.

ESTELLE'S STORY: FROM PANDAS TO CONDORS

"I KNEW I WANTED TO WORK with animals since I was very little," Estelle remembers. In kindergarten, when the class was creating "Me" books about their hopes and dreams, Estelle drew a picture of herself giving a dog a vaccine. She wanted to be a veterinarian.

Estelle and her family always enjoyed animals and the outdoors. Not long after Estelle was born in Philadelphia, the US Army transferred Estelle's dad, a captain, to Oahu, Hawaii. There she and her mom and dad loved exploring the tropical mountains and valleys. When her dad retired from the army to practice law, the family moved to Arizona, and then, when Estelle was ten and her baby sister was only one, to California. But no matter where the family lived, she says, they always got outside to enjoy nature, and "we always went to zoos."

Estelle's family included many pets: Eric the Doberman pincher; Heidi, also a Doberman; Gretchen the German shepherd; Sunny the parakeet; a cockatiel named Chance; a horse named Skywalker. In high school, against her mother's wishes, Estelle brought home the class ball python, Snickers, for the summer. "But soon," she remembers, her mom, a real estate agent, "was walking around the house with the snake draped over her shoulders."

Drawn to its biology program, and its location, near San Diego Zoo and Scripps Institute of Oceanography, Estelle went to college at University of California, San Diego. She started volunteering at the zoo, collecting data on panda behavior—and, not surprisingly, fell in love with the adorable black and white bears. When the zoo's director moved to Atlanta, she followed him there for graduate school.

Soon Estelle found herself moving to the pandas' native China to collect data for her doctorate. Working at the Chendu Research Base of Giant Panda Breeding, she made observations of the bears' behaviors and cataloged their vocal communication. She even got to play with a baby panda named Liang Liang.

While in China, a job opening appeared for a conservation and research coordinator in the Animal Department of the Santa Barbara Zoo. The zoo was known for its protection efforts for endangered Channel Island foxes. But

Estelle and her husband, Brad, share their home with Nira, a curly-coated retriever. Today, Nira's still a youngster, but since the breed is known for its ability to sit patiently for hours in a camouflaged shelter, hiding from the birds being watched, Estelle hopes Nira may one day join her at work, studying condors.

the zoo had recently taken on a new project. "We are closer than any other zoo to condor country," the zoo's new CEO, Rich Block, reasoned, "and we need to be part of the condor recovery program!" He hired Estelle in 2006 and put her talents to work on the birds' behalf.

Back then, most of the wild condor nests in Southern California were failing. A scientist based in San Diego found the nestlings were choking on small bits of trash. Research was urgently needed to find out more. So Estelle switched the subject of her PhD thesis from the breeding behavior of giant pandas to the nesting behavior of California condors. Rich likes to boast that Estelle dumped pandas for condors!

Surprisingly, the cuddly, furry pandas and the scavenging giant vultures have much in common. "Both are umbrella species," explains Estelle, "requiring large tracts of land to be set aside to protect them—land which, at the same time, protects many other plants and animals in the same ecosystem." And both species presented Estelle with similar scientific questions. How long do these long-lived animals stay with their parent(s)? (Giant pandas stay with the mother a year and a half to two years—roughly the same amount of time that condors stay with their parents.) Do youngsters fare better being reared by humans or in the wild by their own kind?

"Before too long," Estelle said, "I was living and breathing condors." Immersed in the condors' world, she's learned how to hide food at her campsite from marauding bears, and how to avoid skirmishes with mountain lions. She's had to drive over a burning log to escape a raging forest fire. She's learned wilderness first aid.

And as well as collecting and interpreting data, her work demands strenuous hikes, carrying twenty-five pounds of supplies and scientific gear into wild, rough terrain. The moun-

Estelle at her desk at Santa Barbara Zoo. On her computer, she is tracking the Global Positioning System (GPS) flight paths of the Southern California condors.

tains are so steep her ears pop as she climbs; and at the end of the long days, her muscles ache. She's sprained her ankles, strained her shoulders, and kept going through hot, dry summers and cold, windy winters.

It's a good thing that Estelle was always physically fit. In high school and college, she worked summers as a lifeguard. She excelled in track and field—even though, she notes, she didn't have a typical runner's long legs and tall, slender frame. In college, she sometimes felt annoyed that her body wasn't the same shape as the other athletes she saw on TV or the models in magazines—but she eventually realized that what's most important about a body is that it be strong and healthy.

Estelle had to be tough and fit to do the work she's always dreamed of doing—the work she continues today. She wants kids to know that "it doesn't matter if the people you see in magazines or on TV doing what you want to do don't look like you." No matter if you're short or tall, thin or not, Estelle will tell you, "Don't let anything stop you from pursuing your dream!"

Condor country.

Chapter Two:

PLAN OF ACTION

IN THE CONFERENCE ROOM at the Hopper Mountain National Wildlife Refuge Complex office in Ventura, California, we can feel the excitement building. Under the watchful red glass eye of a stuffed California condor, fifteen people—including Estelle, photographer Tia Strombeck, and me—gather round the big table for a meeting about one of the most important events of the condor project's year.

"Big things!" announces Joseph Brandt, a tall, strong guy with blond dreadlocks, Pacific blue eyes, and a tattoo of a condor wing on the inside of his right arm. A supervisory wildlife biologist for the US Fish and Wildlife Service (USFWS) and the field crew leader for tomorrow's activities, Joseph calls the meeting to order.

"Tomorrow," he tells us, "we'll go out to Bitter Creek—where there are fourteen condors to handle and process!"

Bitter Creek National Wildlife Refuge, in the foothills of the southwest San Joaquin Valley, is the main release site for zoo-bred condors in Southern California. It's also one of a handful of places where, twice every year, condor conservationists attempt to recapture every wild-living condor. Each individual bird is so precious that the condor team tries to give all of them a health exam twice yearly; they also check and, if needed, update or replace the telemetry transmitters that researchers use to keep track of the birds' movements.

For the people who participate, it's a special time to get up close and personal with one of the rarest and most magnificent birds on Earth.

Tomorrow there will be a lot of big vultures for the team to handle. The enticement of a stillborn calf lures the birds to an outdoor trapping chamber, which then opens up to a large flight pen furnished with another yummy carcass, water, and tall perches. The strategy has worked exceptionally well this fall.

"With fourteen birds to process, if there's anyone from the zoo who can give us a hand, that would be great," Joseph tells Estelle. She agrees to round up some zoo staff for the job.

"On Tuesday and Wednesday, we'll be handling condors. Is there anyone with questions or something to add?"

Tia and I have a million questions. How do you catch a condor in the flight pen? How can you hold on to these strong birds without hurting them? Can we avoid getting bitten by those big beaks, so capable of ripping flesh?

But most of the folks around the room already know the ropes. They know how to grab a bird half the size of a small sofa in their arms, and how to hold its impressive beak, enormous wings, and big, turkey-like feet still

At the planning meeting, from right to left: US Fish and Wildlife biologists Molly Astell and Joseph Brandt, and Santa Barbara Zoo biologists Nadya Seal Faith and Estelle Sandhaus. Sy Montgomery, the author of this book, sits next to Estelle.

for long enough to perform a physical and a telemetry check. Tia and I will learn tomorrow.

"Ok, then," says Joseph. "Now, let's hear your reports."

Everyone here has an essential role to play in following and safeguarding the wild Southern California flock of seventy-six birds. It's a huge job. Each condor may travel 150 miles (241 kilometers) in a single day. The population ranges over an area covering 15,000 square miles (38,850 square kilome-

ters), much of it rugged canyon country. Like Estelle, the workers gathered here have hiked for hours over game trails to observation points, sometimes in one hundred–degree weather, carrying water, camping supplies, and scientific gear. Some, like Joseph, rappel up and down cliffs to reach remote nests— even hauling up an oxygen tank when they check each chick, in case one needs an extra supply to recover from illness or stress.

The folks in this room have a Herculean task—and one that involves duties outside most job descriptions.

"Josh," Joseph asks a bearded and

bespectacled colleague, "how did last week go?"

Josh Felch is a US Fish and Wildlife biologist, and a highlight of his recent work week involved getting the torsos of two dead elk up to Bitter Creek. "They came from San Luis Refuge," he explains. One elk died of a broken neck; the other died during capture for another biologist's study. The torsos, plus twelve elk legs that hunters discarded, will be frozen for later use as trapping bait or to lure birds to a specific site so researchers can regularly check on them.

The condors enjoy but don't need this supplemental food, Joseph explains. There's plenty out there for them to eat. Their usual fare, corpses of wild animals who died natural deaths, are supplemented by the remains of dead farm animals, gut piles left behind by hunters, and the bodies of animals like gophers and coyotes who farmers, ranchers, and landowners shoot because they consider them nuisances. Animals shot by humans provide condors with a bounty they never encountered before the first settlers and their guns appeared in North America.

But that plenty, as it turns out, is the condors' biggest problem.

For decades, the major cause of the condors' decline was a mystery. Were people shooting them? Were they dying from eating

A condor might travel 150 miles (241 kilometers) a day. The Southern California population can range over 15,000 square miles (38,850 square kilometers).

poisoned carcasses? Was their habitat usurped by human homes? Were they getting electrocuted by landing on power lines? Blenderized by windmills?

All these dangers are, in fact, real threats to condors today. "There's no fifteen thousand square miles (nearly thirty-nine thousand square kilometers) that don't have potentially dangerous human activity," explains Joseph. "There are oil fields around Hopper. Condors perch on the pumping units of the oil derricks. There's litter, hardware, nuts and bolts, chemicals. A large portion of condors in our flight area fly near wind turbines."

The US Fish and Wildlife Service has worked with the energy companies to develop best management practices to reduce the dangers. For instance, using the condors' telemetry, wind turbines can be programmed to shut down when a condor approaches. The oil and wind companies "are trying to do the right thing," Joseph says. "They value the condors, too."

But as dangerous as oil fields and wind turbines might seem, these projects aren't the biggest killers. The worst danger, and the cause of well over half of all condor deaths to this day, is lead poisoning from bullets in the carcasses they eat.

The condor doesn't even need to swallow a whole bullet to be lead-poisoned. A bullet leaves a toxic trail as it enters the victim's body. Just a small piece of the cometlike tail the bullet leaves as it travels through an animal's flesh is enough to kill the condor who eats it.

This is one reason the health checks we'll be conducting tomorrow are essential. Any condor who appears to be sick will be taken to Los Angeles Zoo for veterinary treatment. There, lead can be cleansed from the blood (though not from bones or organs) by a process called chelation (pronounced *kee-LAY-shun*). Every captured condor will have blood samples drawn, to be analyzed at a lab for signs of lead exposure. Because so many condors still die of lead poisoning, every single surviving wild condor is a fragile miracle—and that's why scientists try to keep track of each bird's every move.

Around the table, each researcher reports his or her news: We learn that 819, 328, and 895 are hanging out together. Condor 819 was seen feeding. Condors 811 and 818 were

Condor country is pocked with manmade objects that can prove hazardous, from wind turbines to these oil derricks.

observed at Bitter Creek, in an area called Orchard Draw. Number 247 flew to a place called Portal Ridge, and folks think 156 might follow.

Another condor sports a GPS transmitter, but instead of transmitting a hit every five minutes as it should, zoo conservation and science associate Nadya Seal Faith reports she has heard nothing for days. "Probably nesting in an area with no cell service," she suspects. (Though GPS gets its position from orbiting satellites, the data are downloaded from cell towers.) 542 had a telemetry malfunction . . .

To Tia and me, it's all a blur of numbers and names of places we don't know. But to Estelle and the other researchers at the meeting, the places are as well known as their own neighborhoods, and the numbers are as familiar as the names of friends and family. These people know each condor as an individual. They've watched some of them hatch. They've held them in their arms. They've cheered while watching others incubate and raise chicks—and in some cases wept as lead-poisoned condors have died in their hands.

That's why Josh's news garners special excitement. "We have a courtship video!" he reports. A romance is brewing right now, inside the flight pen, where a video camera films the trapped condors who await their health checks tomorrow. Male 20 was seen extending

his wings, bowing his head, and swinging side to side in front of female 654.

Smiles and murmurs of approval erupt around the table. Everyone knows both these birds well. Condor 20 was one of the original condors trapped in the 1980s from the wild. He became part of the captive breeding program and, after fathering many chicks, he was released to the wild again in December 2015. The object of his attentions, Condor 654, is the daughter of dad 247 and mom 79. She fledged (learned to fly) out of a cliffside nest cavity near Hopper Mountain National Wildlife Refuge in 2012. Biologists watched 654's infancy and fledging on a live-streaming nest camera, the first such camera used in the wild for reintroduced condors in the California backcountry. (Now there are three nest cams for Southern California.)

So far, 654 has not shown any interest in

20's overtures. "But at least," says Josh, "they were in the hotel room!"

Every potential condor romance is carefully observed in hopes it will lead to more condor chicks in the wild. Most California condor chicks today are born in zoos, where the eggs and chicks can be shielded from harm; later, youngsters who are not kept for the captive-breeding program are released to the wild. But their babies will face much slimmer odds of survival. Last spring, six wild nests here in Southern California yielded only two surviving chicks. One, a little female, hatched at a place called Devils Gate at Hopper Mountain. Erin Arnold, Santa Barbara Zoo's condor nest biologist, has been watching the nest.

"How did the observations go?" Joseph asks her.

"I was out for four hours," the blond,

ponytailed researcher answers. Using a spotting scope to view the cliffside nest entrance from across a sandstone canyon, Erin saw the chick emerge, and the youngster was visible for a full hour. "She was going to the edge and stretching her wings, then going back in. Fledging could happen any day now," she reports.

The other chick, who hatched at Orchard Draw, has already started to learn to fly. Erin tells us this chick is hanging out with other youngsters who were born at LA Zoo and recently released at Bitter Creek. Tia and I might even catch a glimpse of him while we're out there.

"There are birds up in the Sierras," reports

Steve Kirkland, condor coordinator for all the release sites. Using data gleaned from the birds who have GPS transmitters, he reports on the condors who recently visited this northerly area: Condor 360 made a flight up there. So did the couple courting in the flight pen, 20 and 654—in the company of Condor 796.

More news from the crew: 289 and 374 made flights to Kernville. 374's mate, Condor 79—one of the oldest females in the flock—is in the LA Zoo for cataract surgery right now and will be released in the spring. Condor 816 also went to the zoo, for treatment of a ripped nictitating membrane, which is like an extra eyelid that protects the eye, but one that birds can mostly see through. (An extra eyelid comes in handy when you make your living plunging your head inside carcasses!) She'll rejoin the wild flock soon.

Finally, with everyone reporting in, nearly every bird in the Southern California flock is accounted for. The meeting's at an end. "Cool! Sweet!" says Joseph. "You're all doing a great job. It's lunchtime. I need something to eat!"

Estelle, Tia, and I take off in the Condor 4-Runner, a Toyota SUV painted with perched and soaring condors and the slogan WILD 4 CONDORS. After a stop at a grocery store to pick up food for the next few days, we'll make the winding, mountainous, 80-mile (128-kilometer) drive to Bitter Creek. We hope to arrive in time to set up our tents before dark.

We'll have a big day tomorrow.

THE NAYSAYERS: CAN CONDORS BE SAVED? SHOULD WE EVEN TRY?

BY THE EARLY 1900S, it was clear that California condors were disappearing—but nobody knew why. Worse, some people were glad about it.

"What latest madness is this?" wrote an indignant Mrs. Henrietta Baker in a letter to a newspaper in response to the suggestion that a sanctuary be established to protect the birds. If condors disappeared from the planet, she wrote, "I say fine—we'd be better off if they did become extinct, and the sooner the better." She insisted that the birds were "a menace to young pigs and similar livestock" and expected "some rancher's young child" might be "carried off" next!

Ignorance like Mrs. Baker's was widely shared. Some people insisted condors were killers. Others claimed they spread disease. Many thought them dirty. In reality, the opposite is true. Because condors are not predators but scavengers, they prevent the spread of rot, odors, and disease by cleaning up carcasses. They're exceptionally tidy birds who love to bathe. They spend lots of time carefully cleaning their beaks and heads and grooming their feathers with their beaks.

But even the naturalists and biologists who studied them couldn't solve the biggest mysteries about California condors. They couldn't figure out what was killing them. And they couldn't agree on what to do to help.

The recovery effort for condors was one of the most hotly debated in conservation history. By the late 1940s, some experts argued that the best thing might be to capture some wild condors and begin breeding them in captivity. But others strongly disagreed. The first condor biologist, Carl Koford, argued against it. When the US Fish and Wildlife Service began its research program on wild condors, the Sierra Club accused its biologists of invading condor nests and "deranging" natural breeding. Conservationist David Brower, founder of Friends of the Earth, insisted captive-bred condors would amount to nothing more than "flying pigs."

Today, thanks to the success of efforts that were once widely and strongly opposed, more condors fly free in the wild than live in human care. But even now, condors are still vulnerable. Without further protection from lead, trash, and other hazards, even condors' most optimistic supporters say the wild population remains at risk.

This young condor hasn't yet developed the bright facial coloring it will have in adulthood.

Chapter Three:
CONDOR CHECKUPS

"WE'RE SOON GOING OUT to the flight pen," USFWS wildlife biologist Molly Astell tells the fifteen of us gathered in the morning in front of the Bitter Creek bunkhouse. All seven bunks were full with staff and volunteers, which is why we brought our tents. We spent a chilly night curled tight in our sleeping bags, listening to the songs of coyotes.

Molly has exciting news: There are now even more condors in the flight pen—twenty-two of them, including three older "mentor" birds who will attract other condors to the pen and help keep the younger ones calm. Our day will be even more hectic than we'd imagined.

Having worked with these huge birds for eight years, Molly offers this advice: "Keep your voice low, and if somebody else is loud, feel free to tell them to quiet down. Stay still as you can. If you're moving around, the bird will react—and not in a good way."

"And if you've got jewelry," adds Joseph, "take it off. Josh once got the toe of a bird caught under his wedding ring!"

"Our handle time is fifteen minutes," Molly tells us. During that quarter hour, we'll have a lot to do: Draw blood. Do a physical exam. Take measurements. Check telemetry—and, if necessary, replace it. That's not much time, but we'll work as fast as possible so as not to unduly stress the birds. "The end goal of the day," Molly reminds us, "is to not hurt the condors—and to not hurt ourselves."

We pile into three vehicles for the short drive to the flight pen.

What we find when we get there is breathtaking. For Tia and me, meeting 174 and 603 at the zoo was thrilling enough; the chance to spot a single *wild* condor, even at a distance, would be wondrous. But here, before us, are more condors than most people will see in a

lifetime. And though they've been trapped temporarily, they are all living wild and free.

Most of the huge birds perch on dead tree branches; two pull at the entrails of a calf carcass on the ground. We admire the ruff of black feathers gathered like a cowl around the base of their necks. "It's the only bird that has its own feather boa!" exclaims Tia. A handful of condors pace calmly along the ground. Each looks like a distinguished elderly man, clad in black, striding slowly, thoughtfully, with hands behind his back.

"You are looking at a quarter of the wild Southern California population," Estelle tells us.

Four AmeriCorps volunteers set up a canopy to shade the people and the birds from the blazing sun during the checkups. Even though it's mid-November, it gets hot enough by day for a person to get sunburned, and a stressed

A delicious calf carcass entices condors young and old to the flight pen.

bird could overheat. Meanwhile, other helpers disgorge cases of medical and scientific equipment from the cars. On a folding table, staffers set out new colored and numbered wing tags, telemetry devices, and data sheets. Nadya readies vials and trays for blood draws and feather samples. Erin lays out screwdrivers, hole punchers, batteries.

As Tia and I gaze at the birds in the flight pen, we notice we're not the only ones watching them. On top of the pen, male Condor 480 peers in at his friends. On the ground, another condor, his black head showing that he's less than five years old and not yet mature enough to find a mate, hops closer to investigate the

birds in the pen. He cocks his head at us, eyeing us with wary interest.

"They're so naturally curious as scavengers," says Estelle, "they can't help checking everything out. They're such social birds!" She's seen a condor patiently wait around the pen for ten days until his mate was processed and released. "So sweet!" says Tia. "Yeah, but some others are waiting for different birds to get out so they can fight with them!" adds Estelle.

Molly and Dave Meyer, who's been with the project just two months, now approach the door to the pen with huge nets. The condors have seen this before. As they start flying, the

air fills with the sounds of rustling feathers—like dozens of debutantes in taffeta ballgowns rushing down a flight of stairs. Joseph wields a tall, extendable pole tipped with a tennis ball to help urge birds to move from inconvenient perches.

The team has an eye on 20—the one who was observed courting 654 on video in the flight pen. Obviously he's not particularly upset about being in the pen. To him, the health and telemetry check is old hat. "He's thirty-seven years old and has been handled many times," says Joseph. That makes 20 a perfect demo bird for the newbies in our group.

The three chase 20 across the pen till he flies to a high corner. Molly nets him almost instantly. Molly makes sure 20's pink and orange head, with its sharp beak, is at the far end of the net so she can pick him up safely by his back end. She takes the tail end of the bird out of the net and holds his torso between her legs, giving her control of the body and powerful wings.

"You want to keep in mind where the head's going to be when you get the bird out," Molly stresses to Dave. "Remember, condors are always faster than you, so move slow!" Molly walks out of the flight pen, holding the massive bird's back against her chest with one arm, his sharp beak in her opposite hand.

Joseph takes a seat in a folding chair under the shade of the canopy. Molly, still holding 20 still, sets the bird in Joseph's lap.

"My hand comes up under Molly's grip on the beak," Joseph narrates. He slides his right palm up the underside of 20's neck. He cradles the bird's chin in his right palm while wrapping thumb and forefinger around the long, sharp beak. Rather than let the bird extend his neck across his knees, Joseph gently folds the neck back, closer to the body. That's because the condor's flesh-ripping power comes from pulling the neck back, not shooting it forward. In this position, "you're taking away that pulling power," Joseph explains.

He can testify to that strength—and to the damage a condor's beak can do. Once, while rappelling up a cliff to check on a condor nest, the mother—it was Number 147, he remembers—flew at him unexpectedly and landed on his head! While hanging by a rope hundreds of feet in the air, Joseph shook the angry mother condor off—but she swung around, landed on his back, and bit his arm. He still has the scar.

Another time, Joseph had come to check on an egg that was just about to hatch in a different nest cavity. The egg had been laid by a bird at LA Zoo (it was replaced by a wooden egg, which the parents resumed incubating)

and placed in the empty nest of a very dominant male, Number 107, and his mate. As Joseph was sitting at the entrance to the fifteen-foot-long (four-and-a-half-meter-long) nest cave, he could hear the chick inside the egg. "I must have been paying attention to the egg, so I didn't notice 107 until I looked up and he was right there! He jumped at me, landed in my lap, and bit my chest. I had a hole in my shirt and a scar on my chest."

Number 107, he adds, was an exception. "Condors look big and scary, but they usually tolerate your being there, even in the nest." If a nestling is removed for a health check or for medical treatment, typically the parents don't attack; instead, they wait nearby until the chick is returned.

"They're scavengers," Joseph says. "They're not killers." Usually they are remarkably tolerant and gentle. And besides, they also have a secret weapon: vomit! Like many other vultures, condors can shoot

out their stinky, acidic throw-up like mace or pepper spray. Because nobody wants to go around wearing the stench of vulture vomit, this defense is almost as effective as a skunk's spray. They seldom need to bite. But when a condor does bite, Joseph advises, "You want to clean it out really good. Remember, they have their heads up the anuses of carcasses."

Incredibly, 20 now looks calm and relaxed in the lap of this big, blond human. "Hold the beak pointed up," Joseph demonstrates, holding the big beak shut with only his thumb and forefinger. "Make sure not to touch the

Striding thoughtfully, this adult condor looks like a distinguished gentleman wearing a black coat.

Molly nets her target bird, careful to keep his head (and sharp beak) at the far end of the net.

Joseph demonstrates how to keep captured condors calm during their health and telemetry checks.

eyes or cover the nostrils." He slightly elevates the other fingers, like you might when holding a porcelain teacup by the handle.

The condor's keel—the pointy breastbone where the mighty wing muscles attach—rests between Joseph's thighs. The man's right arm cradles one folded wing, while his torso presses the other one close. "It's a comfortable position for him," Joseph explains. "A lot of condors relax like this if they're handled well."

20 stays calm even as Molly takes his left foot in her hand for blood samples. The foot looks more like a giant chicken's than a raptor's; happily for us, condors don't possess razor-sharp talons, but instead, claws— toenails as long as a big dog's.

Blood yields much information, including the sex of a bird (you can't usually tell by looking at them), levels of stress hormones, and more. Several vials are drawn from each bird. Perhaps most important, each condor's blood will be tested for lead levels. "According to the blood work," Molly says, "about twenty percent of the flock is exposed to lead each year." But lead doesn't last very long in the blood, adds Joseph; it spikes and drops over a three-week period. The blood samples provide only a short glimpse of the bird's most recent lead exposure history. "Our estimate of lead exposure," he says, "was actually much lower than it really is."

So in addition to blood, at least one condor will donate a feather as well; analyzing feathers will show longer-term lead exposure. The team will pick a condor with a working GPS unit so its travels can be matched to the lead levels found in the feather. Then the team will be able to figure out where in its territory the bird is being exposed to lead.

"The source of the lead is in the food they eat," Molly explains. "An animal shot with lead ammunition, hunted, shot by ranchers, or left in gut piles. A lot of what we do at Fish and Wildlife is promote alternative kinds of ammunition—typically copper-based. It doesn't fragment, and it's nowhere near as toxic as lead." California has now outlawed the use of lead bullets. But condors don't just live in California; and since Fish and Wildlife officers can't be everywhere, it is a difficult law to enforce. Other states are mounting education campaigns in hopes that hunters and ranchers will choose alternatives to lead ammunition.

The blood sample drawn, next Nadya turns her attention to telemetry. These birds are so rare and valuable that the researchers try to place two tracking devices on each bird—either two VHF units or one VHF and one GPS. A VHF transmitter can be tracked by a biologist on the ground or in a plane, using an antenna that looks much like the old TV antennas that used to sit atop houses. A GPS satellite transmitter attached to one wing will communicate a bird's position to orbiting satellites and transmit the information every five minutes.

The more expensive transmitter, the GPS unit, looks like a big clip, like the kind that keeps your opened bag of potato chips fresh. Weighing less than two ounces (fifty-seven grams), it's mounted on the bird's numbered wing tag. The tags and transmitters are attached through a single hole punched into a thin fold of skin in front of the bird's wing. The process is like piercing an ear. In this spot, it won't interfere with flight. If the old hole has stretched past a certain size, it is allowed to heal up, and a new hole is created on the other side. 20 needs a new satellite transmitter. He doesn't even flinch.

He needs a new VHF transmitter, too. The old one had been attached to the fourth (from left to right) of his twelve tail feathers. The feather either broke off or fell out, taking the unit with it. To attach the new device, Molly uses the blade of a scalpel to trim around the central shaft, or rachis, of the tail feather. The ten-gram (about a third of an ounce) transmitter weighs only as much as a feather itself. Nadya squishes some superglue along the rachis and also secures the transmitter with a plastic zip tie for good measure.

"Now I'm going to find a growing primary feather on the wing and notch it to use as a marker," explains Nadya. A primary feather is one of the long, stiff feathers at the tip of the wing that fan out like the fingers of a hand to

Drawing a blood sample.

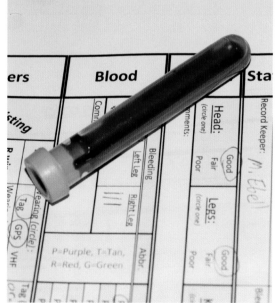

Together with other data, blood samples yield critical information for the condor restoration effort.

25

allow the condor to steer while in flight. Measuring a standard length from the point where the feather emerges from the bird's body, Nadya cuts a little notch. In the spring, when it's time for another health check, the feather will be harvested (they cut the feather with scissors—it doesn't hurt) and re-measured. Each half a centimeter represents about a day of growth. When that feather is later harvested and sent to the lab, researchers will be able to get a history of the bird's lead exposure over the time frame of the feather's growth.

Estelle records everything that the team does to 20 on a data sheet. Next it's time to record the results of the health check: Does the muscle surrounding the keel—the breastbone—feel firm and strong? Are the eyes clear, alert, and responsive? Is the crop—a muscular pouch at the base of a bird's neck where food is stored before heading down through the rest of the digestive tract—empty or full, hard or swollen? Is anything coming out of the bird's nostrils? Are the feet and toenails free of injury and infection? Are there injuries on the face, head, or beak? Joseph goes through the checklist as Estelle keeps track.

"He's in good health," Joseph says of his old friend, and smiles.

In one fluid motion, Joseph rises from his chair, hugs the bird to his chest, and carries him to a scale. Joseph holds 20, steps onto the scale, and their combined weight is recorded. Soon Joseph will weigh himself and subtract his own weight to get 20's. And then Joseph bends over, gently sets the condor's feet on the ground, pulls his arms away from the bird, and steps back. 20 takes three hops and then, with only two wingbeats, he gives himself to the sky. Less than fifteen minutes after the start of the checkup, 20's flying free again.

Only eighteen more condors to go!

JOSEPH AND MOLLY, Nadya and Estelle supervise and encourage while the rest of us take blood, make measurements, record data, and conduct the health checks. Some of us are holding birds for the first time. It's a thrill to be so close. "She's so pretty, I can't handle it!" says one of the helpers, holding number 526, who has bright yellow patches on her pink and orange head. Like people, condors flush brighter with emotion as well as exertion. This condor is panting pretty hard. It prompts Joseph to make an unusual request: "See if her breath smells funky," he says.

A few of us are taken aback. Isn't the breath of a vulture—a creature who eats carcasses—*always* funky?

Taking a feather sample to be tested for lead exposure.

Feather prepared for lab analysis.

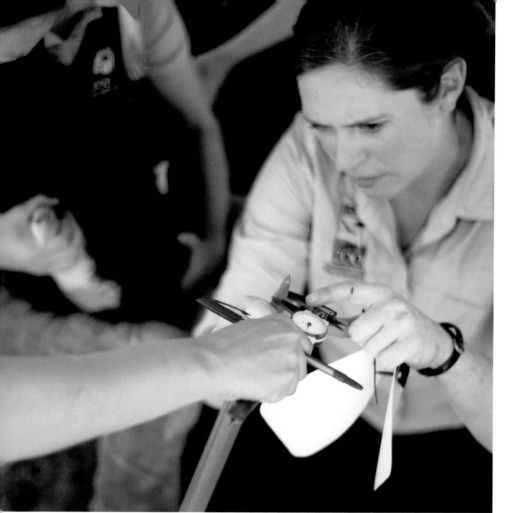

but it could suggest she might have lead poisoning—of which bad breath is one symptom. Because lead paralyzes the muscles, including those of the digestive system, food in a lead-poisoned condor just sits in the crop and rots.

The helper sniffs near 526's open beak. She reports no bad smell. Joseph does his own sniff test to double-check. He carefully feels her crop. "She's alright,"

"I'm not kidding," says Joseph. "See if it smells different from regular condor breath." Surprisingly, healthy condor breath doesn't stink. One researcher described it as smelling like a fresh raw carrot. But bad breath is a sign of a problem.

Joseph is a little worried about 526's panting, especially since her GPS movements showed periods recently when she didn't fly very much. That's not always a sign of trouble,

he announces, to everyone's relief. "She was just stressed. It's OK."

In fact, to our delight, not one of the condors we've examined appears to show any symptoms of lead poisoning, other illness, or injury. Three people got bitten (including me), but more important, no condors were hurt in the handling.

They surely don't enjoy these checkups. But for them, it seems the health and telemetry

exam is probably no worse than a trip to the dentist or taking a math test. Even Number 107—the male who attacked Joseph when he entered his nest—submits without much fuss. "I've watched you for many hours from far away," Estelle says to him softly, as she holds his feet for the blood draw. "You are so handsome! The ladies really like him, Sy," she confides to me. "He's a very successful breeder."

Finally, it's time for the last bird of the day. We've all worked steadily since midmorning, without even a bathroom break, much less lunch. We've saved the final checkup to share with twenty-two zoology students arriving from nearby Taft College, visiting Bitter Creek on a field trip.

Joseph gives a short introduction, telling them how close condors came to extinction and highlighting how far they've come. "In the 1980s," he tells them, "there were twenty-two wild condors left in the world—that's the size of your class! And they all inhabited this portion of California. The last was trapped on this hillside," he says, pointing, "right over there."

Joseph explains that the road to saving condors hasn't been a smooth one. Even the exciting first releases of captive-bred condors into the wild turned out to be a bit of a bust. "Condors are really social, and these condors were just kids with no adult supervision. They

Preparing a feather to receive a VHF transmitter. Next, it is glued and finally tied to the feather securely.

didn't know how to behave in the wild," he tells the students.

Though adults of a related species, Andean condors, were released with them, the juveniles paid their elders no heed. Some approached humans too closely. Others were downright dangerous: one gang of eight juvies got into a house, destroyed a satellite dish, and ripped up a mattress; when the homeowner discovered the mess, one of them was holding a pair of his underwear in its beak! Another young condor visited a campsite, where he found a loaded pistol in a backpack. The condor strolled about the grounds, holding the weapon in his beak by the trigger.

"They were like kids set loose in a grocery store," Joseph tells the students. "That kid's hopped up on sugar because he's eating all the candy. That one's dead because he drank Tide. Everything was wrong with that population!"

When it comes to health checks on condors, many hands make light work. RIGHT TO LEFT: Nadya, Erin, Estelle, Ellie, and Great Basin Institute research associate/intern Laura Echavez.

28

Nadya attaches a new wing tag.

Eventually the captive-bred birds were all recaptured and new rearing techniques were developed, pairing young condors with older same-species mentors, like Condor 174 mentoring Condor 603 at the Santa Barbara Zoo.

Even with more than 250 birds living free and many successfully breeding in the wild, there's more to do. They've been pulled from the brink of extinction, but lead shot is still killing birds, Joseph tells the students. They understand what has to happen to change that.

Dave emerges from the flight pen with a firm grip on male Number 570, the last condor of the day. "He's tired, but he's a biter!" Dave says. As he takes his seat with the bird beneath the canopy, the students murmur and gasp. Few have seen a condor up close. "They're such big birds—and they're bald!" says one young woman with surprise.

Finally, at 4:38 p.m., the last blood sample is taken and the data recorded. Dave sets 570 free. The bird hops once, twice, thrice—and opens his wings. He flies off toward the very hillside where the last wild condor was captured thirty years ago.

Joseph thanks the assembled staff and volunteers: "Good job, everybody, on a successful workup. You were all fantastic!"

"That was a *lot* of birds!" says Estelle. "What a day!"

Joseph releases a condor after the checkup.

TO HOLD A CONDOR

BEFORE THE END OF THE DAY, even Tia and I got to hold birds. Twice I held different condors' scaly, whitish feet for their blood draw. Estelle revealed that condor feet are really black. The reason they look white? "They're covered with dried excrement." I tried not to look alarmed, and I made a mental note to remember to wash my hands before I ate.

"One of our keepers says it's like Purell," Estelle noted cheerfully. Their liquid poop—poop and pee combined, actually, which is the case with all birds, since they have only one opening for both kinds of excretions as well as for laying eggs—actually helps condors keep their legs and feet cool and may even keep them clean. "It's acidic and might even kill germs!" Estelle explained.

I was soon to enjoy the benefits of such purification myself. Because I was holding the feet, I was directly in the line of fire for two somewhat nervous birds. I was pooped on—twice.

You wouldn't think that a stinky carcass would improve by passage through a vulture's digestive system. But it does. I was delighted to discover that condor poop (and my quick-dry pants) smelled faintly to me of fast-food

The author with adult male condor 480.

hamburgers mixed with room freshener.

I felt rather proud of this baptism, and I enjoyed holding the condor's entire body on my lap even more. My first was female 771, whose black head marked her as an adolescent. Birds in this age group are often feisty, Estelle

30

warned me, but as I held her, she remained calm and still, even during her blood draw. When the check was over and I started to rise to get her weight, I feared my grip on her body with my right hand was slipping. I unwisely let go of her beak to free up my left hand.

That's when I got a blood draw of my own. Dave doused the bloody slit in my palm with the disinfectant Betadine. ("Remember, she was just eating a carcass!" Estelle said.) I was the third person to be bitten that day. Good thing there were plenty of Band-Aids!

Later, we captured adult male condor 480, who had earlier been outside the pen, peering in. He rested on my lap as comfortably as a baby. With his keel between my legs, I could feel his heartbeat, fast at first, then slower as he calmed. Perhaps, as I held his right wing closed with my right arm and pressed his left wing against my chest, he could feel my heartbeat, too. I held his neck, head, and beak exactly as Joseph had advised, and the condor never struggled. He moved his head slightly so he could look at my face. His ruby eye showed obvious curiosity, but no fear.

The photographer, Tia Strombeck, and Joseph holding young adult condor 748.

Erin takes notes by the scope while Estelle offers backup with binoculars.

Chapter Four:

BITTER CREEK BABY

IN THE MORNING, AFTER PACKING UP our tents, we drive east and south of camp along a curvy dirt road amid the golden prairie grasses and the leathery-leaved, drought-adapted evergreen trees and bushes in Bitter Creek. Condor country is wild and big. Dense, sometimes impenetrable stands of this thick bush is called chaparral. This is the vegetation you see burning on TV as western firefighters battle to save homes from Santa Ana wind-swept wildfires. Chaparral helps keep condor areas wild; small caves in steep cliffs offer remote nesting cavities.

Here, we're going to look for Condor 895—the Orchard Draw chick we first heard about at the Fish and Wildlife meeting the day before yesterday.

Almost seven months old, 895 is one of only two wild chicks to have survived so far from six nests in all of Southern California.

The other one, the Devils Gate chick, 871, is at a different refuge—Hopper Mountain. She hasn't fledged yet. But 895 has—and as part of a nest-guarding program begun in 2007, it's crucial to check up on him to make sure he's in no danger.

That is, *if* we can see him.

"We might only get a beep on the telemetry," says Erin, "and not a visual. But maybe he'll be good for us and come out!"

The Orchard Draw chick was certainly born to a good family. He's the son of two highly respected adults: thirteen-year-old male 328 and seventeen-year-old female 216. Estelle calls 216 the Queen of Bitter Creek.

"She's a super-dominant female," agrees Erin. Age, sex, and size each play a role in the condor hierarchy, and so does personality—

In the morning we packed up, ready to move.

and everyone recognizes that 216 has what it takes. "When she lands near a carcass, everyone else scatters." She and her mate have been together since 2011 and have tended three nests together before this one. But all their previous attempts to raise a chick had failed.

That's not unusual. "There's a lot of care that goes into getting the chick to survive. They take a long time to hatch, a long time to grow," says Erin. Condors have one of the longest fledging periods (the time between hatching and flying) known to birdkind: about half a year. "And that's a lot of time for something to go wrong," adds Estelle.

Hazards at the nest abound: The egg can be infertile. The eggshell can crack. When parents are out looking for food, a chick can be eaten by predators. A chick can die of lead-poisoned meat delivered by unknowing parents. They can die choking on plastic trash. They can fall out of the nest to plunge to their death. They might be snatched by an eagle.

Number 895 is this pair's first success. So Erin feels especially eager to make sure he's OK.

After a ten-minute drive, we pull over to a parking area. Erin loads the battery for her walkie-talkie—essential for warnings of ever-present fire danger and for communicating with fellow condor staff. She shoulders her backpack. It contains the telemetry antenna and receiver, her spotting scope, tripod, first-aid kit, and more water and food than is necessary for a four-hour observation session. But, she notes, "In this job, you never know where you might end up. You may need to hike somewhere else, to repair a camera, to fix something that's wrong. If we get a mortality signal"—that is, if the telemetry indicates the bird hasn't moved in a long time—"we would try to find him." And that could entail a hike of many hours over steep trails, in hot sun.

But unlike many other nest observation points, which may take hours to reach, the observation post for this nest is only a five-minute walk away. At 9:40 a.m. we reach our destination. From our grassy vantage point we see rolling, golden prairie stretching all around us, punctuated with evergreen oaks. Straight ahead, across a valley, are sandstone boulders and cliffs. The air smells like sage.

"There's a bird!" About one-fifth of a mile (about 300 meters) away in the southern sky, Tia spots a huge black bird, unmistakably a condor. With wings held flat, not in a V like its smaller cousin, the turkey vulture, this condor's flight is smooth as silk, without any wobble, and with hardly any flap. With the naked eye, we can just make out a green wing tag. So that researchers can read them, the numbers on the wing tags are as large as possible, with only the last two digits written; the first number is coded by a color. Green labels a tag as part of the 700 series—which means this bird was born in 2015. The condors in the 100 series wear red tags (for 2009); the other tags are yellow (200; 2010), blue (300; 2011), white (400; 2012), black (500; 2013), and purple (600; 2014). By the time you read this, there will be new colors.

We all search with binoculars, but we still can't make out the number.

Erin sets down her pack and arranges everything she'll need within easy reach: data sheet, clipboard, pencil, water, antenna, receiver unit. She erects the spotting scope on its tripod. The scope will magnify an image to look seventy times closer than it really is. That's much sharper than our binoculars, which make things look ten times closer, though even that is better than the naked eye. Estelle stands beside Erin, her binoculars raised. Somewhere, about half a mile (800 meters) away on the opposite ridge, is the cliff with the nest cavity.

"Is that the nest near that guano?" Estelle asks.

"There's a big, pointy rock with whitewash," says Erin. ("Guano" and "whitewash" are other words for bird excrement.)

The bright white portion of bird droppings is a pasty substance containing uric acid, which birds' digestive systems make instead of pee. So when you see whitewash on some surface, like a rock, it offers a good clue that a bird was perched somewhere above it.

"That black opening in the rock is the nest cavity," Erin points out. "And see, on the other side of the cavity, there are two trees . . ."

"Got it," says Estelle. "There's a big boulder at the top. With the black opening, it looks like a yelling face!"

Because 895 can now fly, he may not be inside the cavity. But he could well be nearby. First, though, Estelle wants to see if she can get an ID on the south-flying three-year-old. "Every time we see a condor we try to get a tag number. It might be the only way we have to know that bird's still alive," says Estelle. We all scan with binoculars but can't find the bird.

Erin turns back to the task of locating the Orchard Draw chick. "Oh, I think I see him!" she says. "It might be 895, because he's in his normal tree . . ." She puts her eye to her scope. Before Tia and I can find any bird at all, Erin reports, "No—it's a hawk. Never mind . . ."

With reddish wings and yellow feet, the Harris's hawk takes off and flies by us.

"Wait—I've got him! I think it's him!" cries Erin. "He's on his perch—a dead branch—in the sun. But I can't see his tag . . .

"He's turned around! Now I see his tag—895! He's hunkered down on that branch! Oh, yay—this is so great!" Erin says. And soon the three of us have 895 in our sights, too.

Erin notes the time—10:05 a.m.—on the data sheet, along with his number, her own name, the date, the location. There's room to note physical condition, feeding sessions if

he's supposed to be doing as a condor. He's figuring out his world."

The chick turns and reaches with his beak to a spot at the base of his tail, then he runs his beak through a wing feather. He's preening. There's an oil gland at the tail base that he uses to keep his feathers shiny, strong, supple, and even somewhat waterproof. Maintaining feathers in top condition is essential for flight. And it might be the most action we observe all day.

"These observations last four hours," explains Erin, "and I just love it. You can sit for hours and hours and days and days." Sometimes nothing happens at all. Sometimes you don't even get to see a condor. But still, it's worth it," says Erin. "I can sit in silence and listen to the other birds."

We hear the short whistle of the phaino-pepla, a black bird with a crest like a cardinal and brown eyes. We listen for the coo of doves. Periodically we hear the croak of a raven. "It's peaceful and relaxing. And then maybe something big happens," says Erin. "I got to watch him take his first flight. I got to watch him fledge! That's what makes it worthwhile. That's your moment. It's so rewarding."

By 10:34 a.m., Erin and Estelle figure they should check to make sure 895's telemetry is working. They erect the antenna, set the

an adult arrives with a meal (the parents may continue to offer their chick food periodically for a year and a half), flights, and other activities.

Even though he's just sitting there on his branch, we're thrilled. We're watching a youngster at a critical turning point in his life.

Up until a few days ago, he stayed in his cave, receiving regular meals and cuddles from his parents. While they were away, there was still the possibility of danger: depending on how remote the cavity, a land predator might kill and eat a chick. But now he's far more vulnerable to attack from the sky. Golden eagles, for instance, may frighten and harass condor fledglings. So can other condors—like older bullies at school.

"That he's right out here, I would say, shows he has more confidence," says Estelle.

Number 895 hasn't moved for several minutes. But we're still riveted. And what we're doing is "super-important," says Erin, "even if it's mostly just watching him sit. We need to know he's doing OK. And sitting there is what

Estelle searches through binoculars.

Erin listens for the VHF signal.

receiver to 895's frequency (the frequencies of all the condors are a closely guarded secret), and, holding the antenna at waist height, point it toward the fledgling. *BOOP! BOOP! BOOP!* The receiver answers authoritatively. "His VHF transmitter's working fine," says Erin.

The fledgling now plops down on his chest on the branch, lying stretched out like a cheetah, lounging in a tree. "He's lying down. Babies get tired!" she comments.

She swings the telemetry in an arc, searching for a signal from mom, 216. "She's not here," Erin reports. Next she dials up dad, 328. *Boop. Boop. Boop,* the receiver replies, though more faintly than the fledgling's. "He's around! We just can't see him."

At 11 a.m., we can see 895 moving. He's

standing on his perch, leaning forward and flopping his wings around. Is he tippy? Is he going to fall?

"I think he's wing-begging!" says Erin. "The chick is begging for food. I can't see the parent. But he can see him and probably hear him . . ."

Six minutes later: "There's mom flying!" Estelle spots 216. "I've never seen feeding at a tree—only at the nest or on the ground. Does that happen?"

"Yeah," replies Erin, "I've seen it."

Mom has landed in a scrub oak across from us, in back of the tree where the chick is perched. "216 is trying to get down to 895," Erin narrates.

The mom disappears behind the leathery

leaves of the scrub oak tree. "Hopefully she'll pop back out from the branches. Sometimes it takes a while for them to find each other," says Erin.

"They're so big, it's sometimes hard for them to maneuver," explains Estelle.

We watch and wait. 895 stops wing-begging. We can't see 216. We note the croak of a raven and a flock of bluebirds streaking through the sky.

While Erin and Estelle keep the chick in view, Tia has been scanning the sky. "Condor to the left!" she cries. "And there's another condor to the right!"

Erin wants to scan for signals from these condors. Are they the parents? Or are they other condors who might harass 895? Estelle

A young condor stretches his wing.

"I see the mother's feathers sticking out of the tree now!" says Estelle.

"Yep, that's 216," Erin confirms.

Now the Orchard Draw chick stretches out both his huge wings. He takes several steps toward the end of the branch. "I think he's going to take off!" Estelle announces. "There we go! We have liftoff!"

I'm so bowled over by this event that at first I can't find the chick again in my binoculars. Where'd he go?

"He's almost at his [nest] cavity," Estelle says. "Just above it. He just sort of blooped over there." The flight was only about a tenth of a mile (200 meters) and took less than a minute, but it's the most exciting thing we've seen today. It takes a while for it to sink in: we just witnessed one of the first successful flights by one of the most critically endangered birds in the world.

Does the Orchard Draw chick feel pleased with himself? He holds his wings out—but not to fly again. He's displaying the mottled grayish patches under his wings—patches that will turn white when he gets older. It almost looks as if he's celebrating. And in a way, perhaps he is. "That's a beautiful sun posture," says Estelle. He's soaking up some rays—as if to say it's finally time, once again, for California condors to have their day in the sun.

now takes the scope while Erin erects the antenna again. She holds it horizontally and swings it in a circle. The signal will be stronger if the bird is flying. Then she tries holding it vertically; in this position the signal will be stronger if the bird is perched. She tries various different channels. But the flying condors are now out of view.

"Ooh, I got a wing stretch," Estelle narrates the chick's movements from behind the scope.

At 11:23 a.m., Erin has identified one of the flying condors. "It's 76. She could be down in the valley. She could be twenty-five miles (forty kilometers) away by now." She tries other channels. She's located another flying condor, male number 262.

AN ENCOUNTER WITH A MOUNTAIN LION

ESTELLE WAS EXHAUSTED.

It was spring, and she was watching a chick at a remote nest, both taking data and ensuring the baby's safety. Between the strenuous, hilly hike and the many hours watching the bird, she'd just put in another twelve-hour day.

Making her way back, alone, to base camp—where she and several other researchers had pitched their tents, stored their gear, and hung their food from a bag in a tree to keep it away from bears—she thought she heard something move in the bushes. She smelled a strong, harsh scent that was vaguely familiar. But she was so tired when she got back to her tent, she gave it little thought. Besides, she knew she had to rise at 4:30 a.m. if she was to get back up to the ridgetop observation point before it got too hot. She fell asleep fast.

It was predawn dark the next day when she started hiking along the dry section of a streambed back toward the nest. She heard something move in the underbrush behind her. "And when I turned with my headlamp," she said, "it revealed two big eyes!"

Then she remembered why the smell the night before rang a bell: Her office at the zoo was near the snow leopard exhibit, and it smelled like this. Like cat pee. "Like a lot of cat pee. Like from a really big cat."

The scent last night had surely come from the animal who now stood just twenty feet (six meters) away from her.

It was a mountain lion.

And standing between the mother's legs were her two cubs.

"The mother cougar was crouched down, looking straight at me," Estelle said. "Her tail was up. I said 'hello' in a very deep voice. I backed up. She darted to the right, and her little cubs disappeared. I wanted her to know I was not completely helpless, so I tossed some rocks in her general direction—not far enough to hit her, of course! I continued toward the nest, but I walked backwards for a quarter of a mile—in a rocky streambed. And when I finally turned around, I put my spare headlamp on the back of my head."

That was a smart idea. These big cats are stalk-and-ambush hunters, attacking from behind. Wearing the headlamp on the back of her head probably suggested to the mother mountain lion that Estelle could see her no matter which way she turned.

Forty-five minutes later, Estelle made it to the nest. "And," she added, "the chick was fine!" She didn't see the mother mountain lion again.

One of the potential challenges of working in condor country.

Chapter Five:

THE DEADLY LEGACY OF LEAD

THEY LOOK LIKE harmless grains of sand.

The ten items that Myra Finkelstein holds up for us, inside clear, sealed vials, are so small we can hardly see them. Each vial is smaller than an old-fashioned film canister. Some of the objects inside are smaller than the head of a pin.

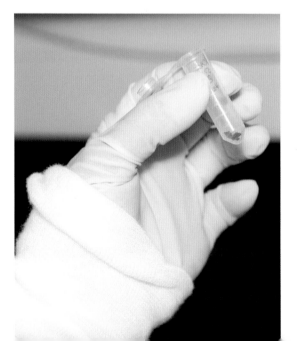

Small but potentially deadly: metal fragments eaten by a condor.

But she and Estelle know how deadly they can be. And that's why Estelle has brought Tia and me here to Myra's environmental toxicology lab at the University of California, Santa Cruz. After leaving Bitter Creek, we spent another night and day at Hopper Mountain, trying—unsuccessfully—to get a glimpse of the year's other wild chick, 871, also known as the Devils Gate chick. We've come from the field to the laboratory today to see how Myra is using science to solve condor murder mysteries.

The tiny fragments in the vials came from Condor 525, a captive-born female who hatched on May 4, 2009 at Portland Zoo in Oregon. She'd been released in November 2010 at Pinnacles National Park in Central California. There she had paired with a mate,

captive-born Condor 405. But last August, her telemetry signals showed she wasn't moving much—a sign she was not well.

Sometimes vets can remove fragments like this through surgery or make the bird throw up the lead pieces. Sometimes they can cleanse a bird's blood of toxins. And, in fact, in spring 2017, 525 had been treated for a month for lead toxicity at Oakland Zoo.

But this time, for 525, it was too late. She was found dead in the field. Her corpse was sent to a federal laboratory. The remains were badly decomposed, but her organs showed very high levels of lead.

"The condor who ate this died," Myra tells us, her green eyes sad. "Just one of these, the size of a few grains of sand, is enough to poison a bird," she says. And since condors are so

At her environmental toxicology lab, Myra Finkelstein supervises as assistant Zeka Glucs prepares a sample.

social, and often share a single carcass, "You can see how one poisoned carcass could wipe out one quarter of California's wild population of condors. It's sobering."

Myra has thousands of condor-related samples in her lab, including more than a hundred fragments like these—some of them removed surgically, some vomited up by the birds, some retrieved from their poop, and some collected after a bird's death in a necropsy (which is what vets call an animal autopsy).

She has analyzed not only lead fragments but also thousands of blood samples. Since the year 2000, she's been testing feathers, too. Over the course of her years of investigation, Myra has proved a pivotal figure in the battle to save California condors. In 2012, her lab's analyses of ammunition and of condor blood samples—like those we took at Bitter Creek—sealed the case against lead bullets that researchers had been working on for years. Myra's team not only showed without a doubt that lead poisoning was the leading cause of death for this endangered species and was preventing its recovery, they also analyzed the chemical "fingerprint" of the lead that was killing the condors and proved it came specifically from ammunition.

Her data led to California's historic law restricting the use of lead bullets—the first and, so far, only such ban in the country, one that Myra and other scientists would like to see expanded to protect other animals all over the country and around the world.

Myra gestures to cabinet after cabinet in her pristine laboratory. "These samples all come from birds showing symptoms of lead poisoning," she explains. "But we need to find the source of the lead. Was it from lead ammunition? From lead paint? If we know what the source is, efforts can be made to remove the source."

She's going to be conducting one of these tests today, and she invites us to watch. She dons her white lab coat, lab goggles, and plastic gloves. Her assistant, graduate student Zeka Glucs, does the same.

"I'm ready when you are," Myra says to Zeka.

Today they will screen metal fragments found inside the bodies of condors to see if they're made of lead. "Each of these samples *could* be mostly made out of a different element. If they are lead," she continues, "we need to put them to more rigorous analysis."

The first step is to rinse the tiny sample with a weak acid. Myra holds it with forceps while she squirts it with the acid, holding it over a tiny plastic tub no bigger across than a packet of sugar. Then she blots the sample dry on a paper towel. Myra picks it up again with the forceps and places it in a small tube. Zeka squirts in more acid and Myra shakes the tube for thirty seconds.

Myra peers at a sample through a microscope.

Rinsing the sample.

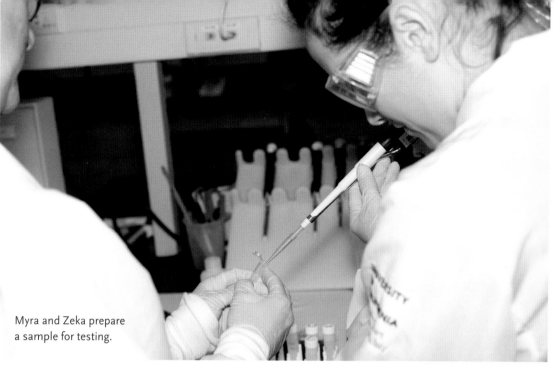
Myra and Zeka prepare a sample for testing.

Myra hands the tube back to Zeka. Using a pipette—it's like an eyedropper—Zeka sucks out the liquid and puts it into a new test tube. The liquid now contains an extract from the solid fragments. The sample is carefully labeled with the bird it came from, a number and letter identifying the sample, and today's date. The original fragment is then transferred to a fresh container for long-term storage and possibly more tests.

The scientists repeat the process for two more samples.

Carrying the liquids in their test tubes, Myra leads our group across a courtyard to the Santa Cruz campus's Earth and Marine Science Building. We're there to use the Flame Atomic Absorption Spectrometer. It's an apparatus the size of a giant Xerox machine with a big hood over it. When an acetylene torch inside the machine burns the sample, Myra explains, this excites the sample's atoms, and they give off a specific wavelength of heat energy—which is absorbed and read by the machine.

"Each chemical element has a different wavelength," she explains. Carbon's wavelength is different from oxygen's, which is different from plutonium's, which is different from sodium's, and so on. This spectrometer can spot them all—but today she programs in the wavelength for lead on the machine's numeric keypad. She sets the sample inside a fat tube and slides it into the spectrometer. A thin tube inside sucks the liquid sample out; the torch vaporizes it; and the machine offers its reading: the first sample is indeed lead.

But where did it come from? "We know California condors—and other wild species—are lead poisoned all the time, around the country, around the world," Myra tells us. "It's very important to understand the sources so we can fix the problem."

Lead may come from many sources. Lead is naturally found in rocks; it was added to gasoline until it was outlawed in 1996; it's in batteries, ceramics, certain kinds of glass, and the lead sinkers fishermen use in states where they're not outlawed. The list goes on and on. And though lead-based paint was banned in the US in 1978, old, peeling paint still poisons people and wildlife.

BEYOND CONDORS:
THE CRISIS THAT CROSSES SPECIES

THE CAMP'S OWNER had been watching the loons on her New Hampshire lake for years. She loved these large, black-and-white-patterned water birds, with their ruby eyes and their eerily beautiful, quavering calls. She knew each one like a neighbor, and she kept track of when they returned to the lake from their wintering grounds on the ocean; where they nested; when they had chicks; and how many hatched each year.

One day, she noticed that one of the adults seemed sluggish. The bird was swimming too close to shore, eyes half closed. When a boat came near, the loon didn't try to get away. It didn't call; it didn't dive. The next morning, she walked outside her camp and found the loon fifty feet from her front door, at the edge of the water, with its majestic black head drooping almost to the ground.

She phoned New Hampshire's nonprofit Loon Preservation Committee (LPC). A staffer rushed over to pick up the sick bird and take it to the Loon Center, where wildlife veterinarian Dr. Mark Pokras was conducting a necropsy on another bird.

By then, the loon could no longer hold its head up. Its eyes were closed. The bird was breathing slowly and noisily.

"I've seen this way too many times," said Mark. A retired professor at Tufts Wildlife Clinic and Center for Conservation Medicine, his research on lead poisoning in loons brought about bans on lead fishing tackle in a number of states.

The Loon Preservation Committee's studies found that poisoning from lead fishing tackle was—and, despite new regulations, still is—the number one cause of death for loons in New Hampshire. Despite new laws, people still use lead tackle, and because lost sinkers, jigs, and split shot sink to pond and lake bottoms, where loons can swallow them, lead is still killing loons. In fact, in New Hampshire, lead poisoning is six times more common than the next most common cause of loon death (which is physical injury, often from colliding with boats).

Of course, the problem with lead is not confined to New Hampshire, or to California, or to loons and condors. In 1991 alone, biologists and conservationists estimated that two million ducks died from swallowing lead. Scientists have documented more than 130 species of animals who are sick or dying around the country because of lead poisoning. They include animals from eagles to swans, deer to bears—and people, too. Some researchers even blame lead poisoning for the fall of the great Roman Empire in 455 AD. (In ancient Rome, water was carried into homes through lead pipes, and people often cooked in pots lined in lead. Some historians believe Roman leaders were so weakened by the toxin, they were easy prey for invaders.)

What makes lead so dangerous? Lead displaces important minerals from the body. The most important of these is calcium. "This mineral is essential for a huge number of reactions in the bodies of people and animals," explained Mark. Everyone knows that calcium is important for building strong bones. But nerves, too, need calcium to send signals around the body. We need it to digest our food and to learn about our world, to think and remember and react.

Young, growing creatures—including human children and babies in the womb—are particularly vulnerable to the ravages of lead. Because they are growing, human children absorb four to five times as much lead as adults would from the same source. Lead can poison the kidneys and liver, too. But the most dramatic damage affects the developing brain. Even very small amounts of lead can cause permanent intellectual disabilities and behavior problems in kids. The federal Centers for Disease Control and Prevention considers no amount of lead in air, water, or food safe for children.

Few animals appear to be immune to lead's poisons. A recent scientific paper from Canadian researchers showed that some earthworms that live in soil at shooting ranges, for instance, appear to be able to store lead in their bodies without succumbing to its toxins. All of that is good for the earthworms, but bad news for the robins that eat them, because lead accumulates in the body. "If a robin would eat that worm," says Mark, "it could be enough to kill the bird."

At the Loon Center, Mark conducted a quick blood test of the sick loon. The lead levels "were off the charts," he said. The only hope was to surgically remove any lead objects and medicate the bird to cleanse its blood at the sophisticated bird rehabilitation center Avian Haven—a three-hour drive north, into Maine. So LPC staff wrapped the loon in a towel, put it in a cardboard box, and with the bird in his car, Mark started driving as fast as he legally could. As he drove, every once in a while he would hear the sound of the sick loon's beak pecking weakly at the cardboard.

Then the pecking stopped.

An x-ray of the loon's body revealed the culprit: a lead fishing jig showed up clearly on the film. But by then it was too late. The loon had died on the way to the bird hospital.

Loons like this one are among the more than 130 species of animals vulnerable to lead's poisonous effects.

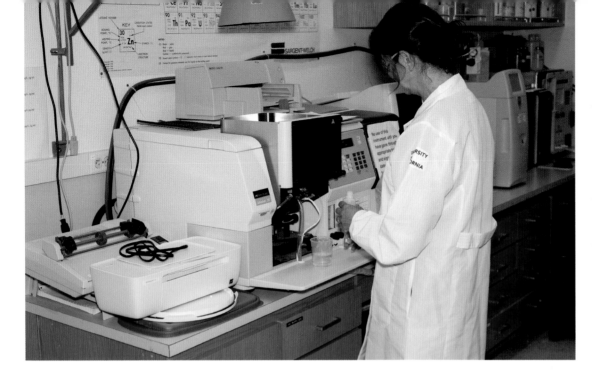

In her earlier work, Myra demonstrated that lead paint peeling off old US military buildings on Midway Atoll in the Pacific Ocean was killing baby Laysan albatrosses—and her data got the government to clean up its act!

That's why the next step is essential. Myra shows us the Inductively Coupled Plasma Mass Spectrometer—ICPMS for short. It's the size of two large freezers and has big hoses snaking out of it.

The machine can measure very low levels of lead, and it can also test for different types, or isotopes, of lead. "There are four different isotopes of lead, and the varying amounts of these different isotopes in a sample can help identify its source," Myra explains. (Different isotopes of the same element have different atomic weights. While all have the same number of protons in the nucleus of each atom, each isotope has a different number of neutrons.) For Myra, finding the specific isotope profile of the lead in a metal fragment, blood, or feather is the forensic evidence she needs. It's the chemical fingerprint of the killer.

In her work on condors, the culprit is almost always the same: though she has found some condors that died from eating lead paint that had peeled off a watchtower, "it's almost always lead from a bullet," she says.

But the bullet fragment that poisoned a bird may be long gone by the time anyone finds the victim. Even if the lead doesn't show up in a blood sample, it can still be deadly.

Blood is cleansed by various organs in the body, including the liver, and blood renews itself. (In humans, for example, red blood cells are replaced every 120 days.)

So as well as analyzing lead fragments and blood samples, Myra is increasingly interested in studying feathers—like the one Estelle brings in today.

"The feathers are so powerful," says Myra. "The blood shows if the bird was exposed during the past couple of weeks. But a feather has a longer history to tell."

A condor's primaries, or flight feathers, grow a fifth of an inch (about half a centimeter) every day—and these "fingertip" feathers can attain a total length of more than a foot and a half (forty-eight centimeters) in 102 days. Therefore, a full-length primary feather records a three-month timeline of a bird's lead-exposure history. One writer called this "a timeline of toxicity."

The feather Estelle brings today, like the others Myra has analyzed, will be subjected to the same process. First the feather is marked, measured, cut in three-quarter-inch (two-centimeter) sections (each section representing a different period of time) and cleaned. Then each section will be dried and weighed. Next, the chemical components of each section will be extracted, and the ICPMS will tell how

Estelle hands Myra a feather for analysis.

much of different types of lead are in each sample—revealing when, at what level, and how often the bird was exposed.

The feather studies revealed a stark reality: "They are poisoned at higher levels and much more frequently than we thought."

On average, according to Myra's feather work, wild-flying condors are eating lead frag-ments *every fifty days*. And more than a third of them are exposed to so much lead that, without human intervention, the poisoned birds would die. Even more die indirectly from the toxin because lead poisoning impairs brain function. As is well documented in children who eat lead paint chips, even very low levels of lead lowers intelligence. Even if the lead levels aren't high enough to kill a condor out-right, it could render the birds less alert and therefore more vulnerable to predators, flying into power lines, getting hit by cars, or miscal-culating a landing.

Condors aren't the only ones at risk from lead bullets: "We know a lot about condors, and that helps us understand the severity of the problem," Myra says. "Hawks and eagles and even mountain lions are lead-poisoned, too. Children can eat game shot with lead ammunition and have elevated lead levels. It doesn't take much. When you realize that no level of lead exposure is safe for a growing child, you see how dangerous this is."

Fortunately, there are alternatives to lead ammunition; there are now bullets made from different combinations of metals such as copper, zinc, tungsten, tin, and bismuth. The most popular of these are copper. They won't poison wildlife, pets, or people. Surveys of hunters show that 90 percent say they like using these new bullets.

They are a little bit more expensive. Copper bullets may cost $10 more than lead for a box of twenty rounds, but bullets are the cheapest part of hunting. A new rifle may cost anywhere from $600 to more than $20,000. A spotting scope can cost even more. Happily, many hunters are also conservationists who want to help protect endangered species. These folks are glad to spend a little more money to take the toxicity out of their sport—especially if their own kids are going to eat the meat they harvest.

But the California ban on lead bullets angered some hunters. They didn't want to change. Certain gun lobby groups oppose the ban and are fighting to prevent expanding it to protect condors and other animals and people in other states. A spokesman for the National Rifle Association even claimed, "The fact is that traditional ammunition does not pose a significant population-level risk for wildlife."

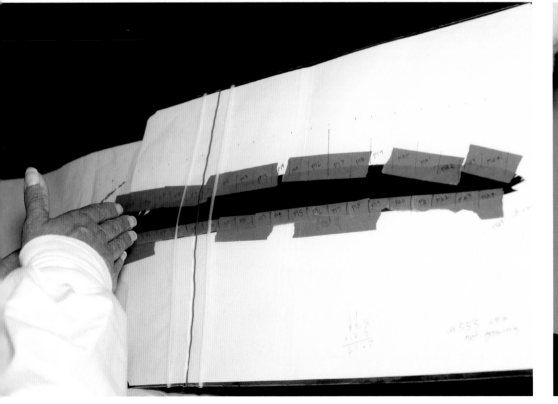

Each segment of the feather represents a different period of time in which the condor may have been exposed to lead.

Myra logs data from her study.

The tiny piece of lead in the photo above, two-tenths of an inch (half a centimeter) long, was poisonous enough to kill a condor.

Science shows his statement is false—and dangerous. In 2013, Myra was one of North America's thirty leading scientists, veterinarians, public health officials, and neurologists who signed a document testifying to lead bullets' scientifically proven dangers. They cited twenty-six scientific journal articles as references. The group all agreed the dangers of lead were so severe, and so clearly proven, that lead ammunition should be exchanged for alternatives such as copper-based ammunition everywhere.

That was way back in 2013. Sometimes it takes a long time for science to prompt change. When Myra first started studying the Laysan albatrosses on Midway Island, it was 1997. She published her evidence, showing why ten thousand birds a year were dying, in 2003. It took until 2017 for the federal government to clean up the site. When it finally happened, she said, "It was amazing. I couldn't believe it!" Finally, her work had won a big victory.

Myra is in this for the long haul. She has new studies planned, like examining how lead affects condors' bones. It's known to alter bone structure and might make them more vulnerable to fractures. She's investigating the effects of different chemicals on their eggshells too.

She's doing the science for the condors. And as the mother of twin seven-year-old boys, she's doing it for them, too. "This is stuff I believe in," she tells us. "You do this work because you love it, because you care. Sometimes you do twice as much work to get the right answer, the best science, the best answer. Sometimes you have to repeat experiments and put in way more time than you thought. But you do it. You do it because you care."

REARING, RESCUE, AND REHAB AT LOS ANGELES ZOO

THE INSIDE OF THIS Los Angeles office looks rather like a miniature spaceflight command center.

Mike Clark and Jenny Schmidt man the four computer screens. They can zoom in for a closer look at the action, or use a joystick to scan a wider area. But instead of watching rockets launching, they're keeping tabs on condor chicks in nest boxes. Like a rocket launch, each chick's infancy is a new beginning; and also like a rocket launch, these babies' trajectory will lead to the sky.

But there's another difference between this scene and a typical command center: a seven-year-old female condor named Dolly perches on a stand in the middle of the room, overseeing the operations. Rescued as a five-month-old from her cliffy nest in Pinnacles National Park after biologists spotted her broken right wing, not even two surgeries could restore her ability to fly. Now, as the world's first official California condor ambassador, Dolly attends fundraising galas and visits with children—when she's not overseeing office work.

Dolly doesn't actually have a say in what happens at the chick cam command center. But here at the Los Angeles Zoo, the staff will

try almost anything—no matter how unconventional—to save California condors from extinction.

The Los Angeles Zoo handles all the veterinary care for the Southern California condor population. They've treated hundreds of cases of lead poisoning. The usual treatment, chelation therapy, starts with five days of injections of medications that bind to the lead so it can be eliminated from the body. But the therapy itself can be toxic. On the fifth day, the condor gets another blood test; if the lead level still isn't below a safe threshold, the treatment begins again after a two-day rest. At times, the zoo staff has treated fifteen condors a day.

They've fixed broken wings and fractured legs and operated on infants to remove bits of trash the chicks have swallowed. But they've also tried some radical new ideas: They've raised baby condors using leather hand puppets as stand-ins for absent parents. They've swapped out infertile, wild-laid eggs with fertile ones laid at the zoo. They've taken fertile eggs from wild nests to hatch in incubators and raise the chicks at the zoo. The parents, living in a world they surely understand is full of predators and pitfalls, waste no time mourning the lost egg; instead, they produce another egg to incubate and raise in the wild.

Zoo staff has captive foster parents raising

Dolly oversees the condor chick cam command center at LA Zoo.

wild condors' chicks, and captive mentors teaching chicks how to be wild. They've even helped a single condor mother do what no wild condor has done before: raise two babies in a single nesting season—a feat that would normally take nearly four years to accomplish in the wild.

"Now, we know what we need," says Mike.

But that wasn't always so.

"When this thing started"—back in 1987, when there were only twenty-two birds left and the roundup began to bring them all to LA Zoo and to the San Diego Safari Park—"it was a rescue mission," Mike tells us as he leads Estelle, Tia, and me on a tour of the facility. "We didn't even know why they were dying.

It was a Hail Mary—a last-ditch effort.

"All this was supposed to be temporary," he tells us, pointing to a series of trailers that house the condor project's offices, nesting chambers, and incubators. The plan was to quickly restore condors to the wild and then

Mike observes a nestling remotely on a chick cam.

get rid of all the offices and equipment. That was thirty years ago. "Now we've learned a lot—and we're still learning. And we're doing stuff we've never done before."

Not everything they tried worked, of course. After all, the first captive-born, puppet-raised condors who were released into the wild proved so unruly that they all had to be recaptured. By 1994, five of the thirteen young condors released had died; four of them electrocuted by power lines. They didn't know how to live as wild condors.

The zoo adjusted its protocol. The eight survivors were recaptured. A mock power pole was erected in their flight pen that delivered a mild electric shock when a bird alighted on it. Mike was astonished to see that young condors didn't have to experience the shock to learn the power poles were dangerous; they learned just from watching other condors. "They're very curious and social," Mike says. "Condors are extremely aware. They're so perceptive."

Learning from parents and other adults is far more important to this smart and long-lived species than anyone expected. Now, at the zoo, most chicks are raised by their parents, not by hand puppets; and the few chicks who are still raised by puppets can look out of their nests to see live adult condors. Around the age of five months, the chicks move from the nest to live with these older mentor condors.

As well as refining earlier techniques, the condor staff have developed new ones. One recent breeding breakthrough involves a love story between Mike and a thirty-four-year-old female condor named Anyapa.

"She's a special bird," says Mike. Hatched at the San Diego Zoo Safari Park in 1984, Anyapa was hand-raised and too used to people to be set free. But at the Los Angeles Zoo, she became an experienced mother, contributing her own eggs as well as raising chicks hatched from the eggs of other condors.

In 2018, Anyapa laid an egg of her own. But in the zoo's incubators, "we had an extra egg," Mike explains. It had been destined for the nest of a wild, childless, condor couple, but bad weather intervened. Mike placed the second egg with Anyapa as well.

In the wild, a single parent raising one chick is unusual. But could a single female raise two chicks alone?

But Anyapa wasn't alone. Anyapa had Mike. Ever since he had cared for her after her left eye was removed in a cancer operation, she had considered him her mate.

Mike could go into Anyapa's breeding chamber and tenderly stroke her head. He could handle her eggs. He could pick up both fluffy white newborn chicks and examine them. He could check their crops to make sure both were eating, turn them over to examine their underparts, touch their stubby wings. Anyapa never considered this an invasion. She considered Mike her partner. And thanks to this partnership, Anyapa was able to raise two chicks in the same nest—making history for the condor restoration effort.

Mike has been a condor keeper since 1989 and is now the zoo's condor program manager. It's been an unexpectedly long and sometimes difficult journey. "We've had birds die in our arms—birds we saw hatch. Birds who we've visited at the nest," he confides. "You need to be strong."

But he has also learned to be hopeful: "The lead thing, just a few years ago, it seemed hopeless." He remembers how, in 2013, the LA Zoo treated fifty condors for lead poisoning. Not all of them survived.

But that was the same year that California passed the first law in the nation to phase out the use of lead ammunition for killing wild animals statewide. And in 2017, only seven condors came in with lead poisoning. "Now look," Mike says, "it's changing right in front of our eyes!"

FIRE ON THE MOUNTAIN

A LOT HAS HAPPENED SINCE November 2018.

Just shortly after Tia and I returned from our autumn expedition with Estelle, we watched in helpless horror as TV and internet video showed the places we had just left being ravaged by what was then the most massive wildfire in modern California history.

The Thomas Fire, so-called because it began just south of Thomas Aquinas College, started on December 4. Driven by powerful Santa Ana winds, it quickly burned through Ventura and Santa Barbara counties. Flames swept through the rugged chaparral of condor country at speeds that sometimes reached an acre per second. The fire raged for forty days, killed two people, destroyed more than a thousand buildings, and scorched 440 square miles before it was finally contained on January 12.

At the Santa Barbara Zoo, Estelle and the staff prepared to evacuate the animals. A handful of creatures—including some reindeer who were visiting for Christmas and a baby giant anteater—were sent to alternative locations in advance. Four precious condors, two griffon vultures, and Veronica the turkey vulture went to the Los Angeles Zoo. As the flames approached Santa Barbara, just about

every animal who could fit into a crate or a carrier was loaded into the Discovery Pavilion. The elephants and a giraffe would stay behind in their concrete barns, defended by firemen. Every zoo vehicle, including the catering trucks, stood by in the zoo parking lot, ready to mobilize once the director of animal care made the call.

At the last minute, it was determined that the flames were not heading into the city. The zoo evacuation was called off. But condor country was still on fire—and the blaze was headed directly for the nest site of the Devils Gate chick at Hopper Mountain. The Orchard Draw chick we had met at Bitter Creek was safe. But for Condor 871, the only other wild-hatched chick to survive to fledge, the situation was dire.

All Tia and I could do was send Estelle our best wishes on email and pray for good news.

Condor staff were evacuated from Hopper Mountain. "We could get signals on VHF on the chick for a week or two from designated safe areas," Estelle said. Then the signals for 871 stopped.

"She's at such a vulnerable age," Estelle emailed us on December 21. "She'd be trying to take some of her first flights with this giant

The Thomas Fire's destructive winter flames give way to spring wildflowers.

fire bearing down on her." Happily, 871's mom, 513, had been fitted with a satellite transmitter, and it was sending continuing hits from the area. "Possibly the chick has fledged to a safer spot on the other side of the ridge," Estelle wrote. "We're eager to regain access to the area to check on 871. Fingers, toes, and primary feathers crossed!"

On December 27, Estelle got the news: When they were allowed back near the site, Nadya and Joseph picked up a signal from 871's VHF transmitter. Finally, on January 2 the team was allowed to get close enough to the nest site to see it. The whole canyon still smelled like smoke. The researchers' boots kicked up small clouds of ash with every step. They could see the fire had burned almost up to the very entrance of the nest. Not far away, Nadya and Joseph found the chick's parents, perched on a rocky ledge above the river.

The two researchers scrambled over a boulder. One hundred yards (ninety-one meters) above the canyon floor, they saw a black spot. It was 871—nearly as big as her parents, and looking strong and healthy. Through their binoculars, they could see her primary feathers had been singed by the flames.

Erin, Sy, and volunteer Kathy Geshweng begin the
long hike to the Snag Ridge observation point.

Chapter Six:
NEST WATCH AT HOPPER MOUNTAIN

ON THIS BRIGHT, CLEAR, BLUE-SKY DAY in June, the smoke and char of the Thomas Fire seem like a nightmare from long ago. Estelle, Tia, and I have returned to Hopper Mountain National Wildlife Refuge. We see few scars from December's flames. Instead, now wildflowers blaze around us, and the pointy, annoying seedheads of foxtail grass are working their way into our hiking boots and socks. It's the season of renewal—and we've come back to see this spring's new condor chicks.

A record twelve wild condor pairs nested this year in Southern California; no fewer than eight of them did so at Hopper. A whiteboard on the wall at the ranch, the Hopper condor crew's field headquarters, tracks the success of each nest.

Unfortunately, at a site called Lockwood,

the egg belonging to male Condor 247 and his mate, 156, proved infertile; at another site, called Pole Canyon, 237 and 563's egg hatched but failed. Nobody knows what happened at the Aqua Blanca nest site; an egg was laid on February 19, but neither parent has since been observed entering the nest.

Nest failures are common. Even in their high, isolated, cave nests, spring's fluffy, flightless condor chicks are pathetically vulnerable. A chick might even suffocate in its egg; it can take three days for a baby to hatch out, even with the parents' help. In nature, a 50 percent failure rate might be common, Estelle says.

But condors no longer exist in a purely natural world. Adults and chicks alike are at risk for lead poisoning from bullet fragments in their food. But the chicks are even more en-

dangered by the tide of human garbage—small bits of trash like metal and plastic bottlecaps, shards of glass, nails . . . anything that might capture the interest of a curious adult who is used to bringing bits of bone—which can be easily digested—back to its babies.

No one is sure why the parents bring these items to the nest. Maybe they're toys. Bones might provide useful calcium and other nutrients for the babies. But because chicks explore the world with their busy beaks, they often swallow these small bits of garbage—called microtrash—which can get stuck in their throats or digestive tracts and kill them.

Humans can intervene, if they reach the baby in time. Using climbing ropes, condor staff can rescue chicks in distress, bring them to the LA Zoo for veterinary treatment, and then reunite the chicks with their parents when the danger is past. That's why it's so important to keep tabs on every chick in every nest—the condor project aims to log at least two hours of observation at every nest at least three days a week. And that's why today we're headed to a nest site called Hopper Canyon, where dad 509 and mom 161's egg, laid on February 18, hatched around April 16.

"You already know the dad, Sy," jokes Estelle. I shake my head—I don't remember handling him at Bitter Creek or seeing him here at Hopper. Then Estelle explains: "He's the condor painted on the side of our car!"

We're heading to a special spot. "The Hopper Canyon nest is at the confluence of several territories," Estelle explains. "It's a great area to check on a couple of nests. We used to come down and see who was roosting for the night. This is condor central!" And the chick, should we see it, will be adorable: at this age the baby will be a gray fuzzball with an orange head. The down won't be replaced with feathers for another month; the head will turn black at four months, and then orange and pink again at five or six years.

"But whether you see the parents or chick is a roll of the dice," Estelle warns. When the chick is small, the parents often swoop in to feed and cuddle it and protect it from preda-

tors. After the chick is a month old—and this chick is two months old—parents' attendance at the nest drops off dramatically. The young chicks spend most of their time where they're safest, as deep as possible inside the cave—where, unfortunately for us, they're hidden from outside view. "The chick is spending a lot of time in there alone, doing the job of growing," Estelle explains.

At nine a.m., we meet Erin and a vol-

Small bits of trash endanger baby condors.

unteer, Kathy Geshweng, a neuroscientist at UCLA, at a spot about a half-hour's drive from Hopper Ranch called Condor Ridge. From the pullover here, it's a steep, hot hike down to the Snag Ridge observation point—an elevation change of 1,248 feet (380 meters). Estelle used to run up this trail. But today, the hike is too steep for her ankle, which is recovering from its seventh sprain. We walk together for a little way down the grassy path before Estelle and Tia turn around to return to the ranch.

Erin, Kathy, and I head downhill along a mown path, surrounded by the tassels of oat grass and foxtail. To keep from slipping down the steep trail, we're sometimes hiking with our feet perpendicular to our path. Along with her pack filled with the usual water, scientific equipment, and emergency supplies, Erin always brings two walking sticks for this trail. She generously lends one to me.

Condor country sprawls around us. We pass rocky Koford's Ridge, named after the first scientist to study California condors. To our right, through binoculars, we can see the towering spires of the Pinnacles, named after Pinnacles National Park, one of the original release sites for captive-raised condors. Beyond Pinnacles we see Hutton's Bowl, among whose canyon walls condors 374 and 289 hatched a chick on April 6. Below us, to our right, a flight pen for capturing condors, like the one at Bitter Creek, comes into view.

But Erin's not looking at the scenery. She scans the sky. "There's a condor!" The tag is black . . . or is it purple? Could it be 616? Then, "There's another one!"

We've already seen two condors and we're not even a quarter of the way to the observation point yet!

Because the slope is slippery and steep, I watch the ground more than the sky as I walk. But here, too, splendor abounds. The silky white mariposa lilies are named for the Spanish word for butterfly. Dense clusters of purple blossoms nod on the long stems of blue dick. Two-inch-wide blooms, bright as daffodils, crown the dwarfish, tough-leaved shrubs known as bush poppy.

By ten a.m., halfway to our destination, Erin takes us a few steps off our trail to remind us that condor country is full of danger as well as beauty. Ahead is a patch of bare ground, not much larger than a backyard deck. It's kept clear of all vegetation so it won't burn. "Here's where you want to get to if there's a fire. The fire shelter is under this bush here," Erin tells Kathy.

I expect to see some kind of bunker, but no. "The fire shelter looks like tin foil and you're a burrito," Erin explains. In fact, the shelter is made of aluminum foil, woven silica, and fiberglass and is only about the size of a sleeping bag—or a coffin. "You'll be hot, and there will be smoke, but at least you are safe from the flames." There are three such sets of gear, as people often travel here in pairs; in less-trafficked areas, like Tom's Canyon, there's only one.

A few minutes later, we pass the path leading to Koford's Ridge observation point. We're now entering a greener zone, one with shrubs and trees, offering blessed shade from the blazing-hot sun. But we shouldn't get too close. Erin offers us some more advice: "Be aware of poison oak. There's bushes of it!" When I took a shower at the ranch, I noticed a quart bottle of special liquid soap that removes the toxic poison oak oils from your skin. Soon we're seeing poison oak almost everywhere.

Poison oak can be the least of a condor researcher's worries in these canyons. Rattlesnakes may lounge, hidden, near the paths. Ticks can swarm by the hundreds. Though I can already tell that our hike back uphill is going to be strenuous, this trail is tame compared with some of the others Estelle and her fellow scientists have hiked.

One of her predecessors, Sanford Wilbur, spent twelve years hiking the steep slopes of condor country, through parching heat, cold winds, thin air, and suffocating pollen. "By the time you reach your destination," he wrote, "you hardly have the energy or interest left to do what you came to do."

Ten thirty a.m.: "Yay! We're here!" cries Erin. In front of us yawns Hopper Canyon, and beyond it, the faces of sandstone cliffs. Past them, fold after fold of mountains, clad in patches of green chaparral, stretch for miles.

We're grateful to set down our packs and

Wildflowers and butterflies adorn our trail.

eager to get started. Kathy and Erin erect their spotting scopes. Beside them, within easy reach, they keep all they'll need for the next several hours. "You want to take your time setting up the scope, because you'll be there, eye to scope, for two hours. You want to be comfortable. You want your water and your sandwich, your pen and your observation data sheet right there. You don't want to miss anything!" Erin advises.

At 10:45, as Kathy begins her observations on the scope, Erin assembles the telemetry antenna. *Boop, boop, boop, bip! Boop, boop, boop, bip!* "Oh, good!" says Erin. "That's 509, the dad! I don't think he's in the cavity. Maybe he might fly in and do a feeding!" Next she dials up the channel for mom, Number 161. No signal.

Erin settles down next to me in a small patch of shade. "I'll let Kathy do the observations now to get some experience," she says. Next time, Kathy might be able to do this alone. As Kathy focuses on the cliff opening, Erin and I scan with binoculars. It's a good time for Erin to fill us in on the drama behind this nest. It's a scandalous condor love story!

"This is 509's third attempt at nesting and 161's tenth," she begins. But this is their first nesting attempt with each other.

Male 509, a nine-year-old who hatched wild here at Hopper, had been the faithful partner of another female, twenty-four-year-old 111. The two had been paired since 2014. In early 2015, 111 laid the couple's first egg. Sadly, their chick died right before fledging. The couple produced a second egg in 2016, but it went missing. Nobody knows what happened to it, but researchers quickly replaced it with an egg laid by captive condors at the Los Angeles Zoo that was just about to hatch.

The couple lovingly raised the chick as their own—and the entire process was streamed live to an appreciative worldwide audience, thanks to the first of two California condor nest cams installed that spring. As shown on camera, 509 and 111 proved to be excellent parents.

Meanwhile, female 161 was happily paired with her longtime mate, 107. During their thirteen years together, the two had successfully fledged six offspring—four of whom grew up at the very nest we're watching now.

All was well with both condor couples.

And then 107 died.

Erin found his body, but it was too badly scavenged to tell what had killed him. He was only twenty-three years old.

Female 161 was left without a mate. But the newly widowed female didn't waste much time mourning, apparently. "She just swooped in and stole 111's mate!" Erin tells us.

What did 111 do about it? "She didn't have time to do anything!" Erin answers. "On GPS, you could see her following the couple around, as if she couldn't understand what had happened."

Why would 509 just dump his previous nesting partner and fly off with 161? Kathy and I want to know. Was there anything wrong with 111? "No—she's awesome!" Erin assures us. "Don't worry—she'll find another mate." Like her rival, 161, 111 was a very successful breeder.

So why did 509 choose 161? We three wonder: Did one have a pinker head? A more graceful flight? Better taste in carcasses? We debate the possibilities. Our conclusion is the same one that other people have reached about many different couples since the dawn of discussion: Love is strange.

By 11:15 a.m., there's still no sign of any of the condors we're talking about. So, as we stare through our scopes and binoculars, we share our own stories. Kathy, tan and fit with her golden-brown hair in a bun, tells us that she's been one of the condor program's thirty volunteers since February. She became an avid birder about two years ago, and found out about the condors through the project's Facebook page, The Condor Cave. "They're such long-lived, social birds, and that's what I love about them," she tells us. "When I saw an ad for volunteers, I said, Oh my God, YES!"

Now, every three weeks she makes the two-hour drive to Hopper Mountain from her home in Los Angeles. Soon she'll be done with her training and able to observe and record on her own for the project. She looks forward to it. "They're fascinating, really unique birds,"

she says, "and I'm grateful to be a part of it and help out in some way."

Erin shares how her path brought her to the condors—from her childhood home in the woods of New Hampshire, to the prairies of Colorado, to the forests of Idaho. She studied bubonic plague in prairie dogs in a genetics lab; she researched the flights of barn owls across a busy highway in an effort to stop roadkills. She learned to rappel by rope into golden eagle nests to monitor infestations of Mexican chicken bugs. These swarming parasites so pester a chick that the baby bird will jump out of the nest to its death to escape them. Before joining the zoo as a condor nest biologist, she traveled to Spain to count birds migrating over the Strait of Gilbraltar . . .

Our conversation is interrupted by another voice. "Oh! That's a northern parula!" Without even casting a glance at the small bluish gray warbler calling from a dead tree, Kathy recognizes who's singing that rising, buzzy trill.

As much as I enjoy my friends' stories, I love to listen to the bird voices around us. We note the onk-a-ree of a red-winged blackbird; the harsh, scolding call of the little brown Bewick's wren.

Condors, by contrast, seldom say anything; they lack the voice box, called a syrinx,

that allows other birds to sing. Though the sounds of their feathers rustling can be heard half a mile away, adults are otherwise mostly quiet. If they're arguing with a fellow condor, they hiss and snort. A chick communicates with its parents through hisses, wheezes, and grunts.

How we'd love to see our chick with its parents right now! At 12:40 p.m., there are now only five minutes left to our two-hour observation session. "C'mon, little buddy!" calls Erin. "Come out, come out!"

But as a pitiless sun blasts us with heat that now has reached a scorching 95 degrees Fahrenheit (35 degrees Celsius), the Hopper Canyon chick sensibly stays safe and cool inside his cave.

"No chick," Kathy says with disappointment at 12:45 p.m. "Bummer!"

"More often than not, this is how an observation session goes," says Erin. "You don't see anything. But you never know when you're going to!"

Erin scans with the telemetry one last time for signals from the parents—or any other condors in the area.

Finding none, we put everything back into our packs and begin the hot, steep, bright hike back up to the top.

Chapter Seven

THE CHUMASH AND THEIR SPIRIT HELPER

WHEN YOU STEP INSIDE the door of the Tribal Council Hall of the Santa Ynez Band of Chumash Indians, you are immediately faced with a heart-pounding sight: a full-size condor in flight.

It's not a sculpture or a painting. It's AC-8, the last wild female California condor to be captured in 1986. Twelve years and many captive-born chicks later, AC-8 was released back to the wild. She flew free, among many of her children and grandchildren, until February 2003. As she perched in a tree, she was shot to death, directly through her wing tag, by a twenty-three-year-old man—an act for which the poacher was fined but not jailed.

AC-8 would probably be alive and perhaps even still breeding today had she not been shot. But her spirit is still alive. Her taxider-mied body serves as the symbol of an undying connection between a great bird and a proud people.

"Our people celebrated Condor," tribal educator Jacy Romero explains to me as we sit together in a conference room in the Tribal Council Hall. Jacy, like AC-8, is a grand-mother, though she looks young, with her radiant smile, smooth skin, and dark hair. "It was important not only for its size, but Condor had special powers to people long ago, and even today."

As one of seven siblings, Jacy learned from her elders about the powers of different wild animals and the connections they had with her people. "The condor is a spirit helper for the Chumash," she tells me. "He could fly long distances. He could foresee the future. In the dance, the condor healed people. A spiritual leader still uses condor medicine for prayers. It gives strength. It gives hope. It gives inspi-ration. It gives long life, wisdom, and protec-tion."

The relationship between condors and the Chumash goes back thirteen thousand years. Rock shelters near Santa Barbara preserve an-cient red, white, and black artwork depicting condors that was created by Jacy's ancestors.

But as Jacy was growing up, few of her people ever saw Condor flying overhead, as they used to when the elders were young. Though spiritual leaders were allowed to own condor feathers, Jacy had never seen any of them don the sacred condor regalia. She had never seen the Condor Dance. The Chumash's spirit helper, it seemed, had been driven from

Condor AC-8 with tribal educator Jacy Romero.

This ancient pictograph depicts a condor taking off, painted over a carving of a bear's footprint.

their skies. Where was Condor? Why was he gone?

At the same time, Chumash culture and language were also under assault. Like the condors, the Chumash once numbered in the tens of thousands. Their territory once stretched over seven thousand square miles (over eighteen thousand square kilometers), from what is now the northern boundary at San Luis Obispo to the western edge of the San Joaquin Valley and stretching along two hundred miles (322 square kilometers) of California Coast. Today there are only about three hundred Chumash on Jacy's reservation; several hundred more live off the reservation.

Once the Spanish arrived in California, many Chumash died of diseases brought by the white invaders. Priests seeking to convert the native people to Christianity established missions and pressured the native people to adopt Anglo culture. Once the Chumash left the missions and returned to the land they knew, they scraped by in difficult times. Because of discrimination, many Chumash stayed quiet about their culture. Only the elders spoke the Chumash language. But they kept their culture alive.

Despite their many difficulties, when Jacy was growing up, her family "had a passion to bring culture back to the people." And as a

result of their efforts, something that seemed like a miracle began to happen: "As we started to rediscover our Indian lifeways," she tells me, "my family would meet with the people from United States Fish and Wildlife Service and zoos that were caring for the condor.

"Lots of questions were being asked," Jacy says. "What was happening with the condor? What was the relationship with the condor and our people?" Her family saw that the condor's fate was entwined with her people's. "Everything started to connect the dots, like the stars in the constellation Cassiopeia," Jacy says.

Cassiopeia is the constellation named

after a beautiful but arrogant woman in Greek mythology; but when the Chumash look at the same stars, they see something else: they see the wing of a condor!

Jacy's family learned that although all the wild condors, like AC-8, had been removed from the wild, they weren't gone. They were breeding at the Los Angeles Zoo and other facilities. One day, they would be wild again.

And as a teenager, Jacy, along with ten other Chumash people from her reservation, were there to see one of those condors set free.

It was a cold and windy day. Her people came with rattles, medicine pouches, and flutes. "We were part of the group of many people working to save the condor," she explains. "And when that bird was set free, you could feel the calmness . . ."

It was not long afterward that Jacy saw the Chumash Condor Dance for the first time. Her relative, a man she had known all her life, donned the sacred costume: hundreds of condor feathers shaped into wings and a skirt. His naked underarms were painted white, like the birds'. Though it was two decades ago, the memory is still vivid in Jacy's mind.

"It was quiet, like a prayer; a quiet time of preparation. There was no drumming. There was no singing. Just a flute and sage smoke. It was heavy medicine."

In that moment, her relative "wasn't just a dancer. People look at that magnificent wing-span, you see that wing going up and down, and it's no longer a person. They took on the spirit of the condor. It was very empowering.

"The dancer wanted to educate the world so much about this magnificent bird. He danced until he was breathless. The dancer transformed." She pauses and lets the magic sink in, then says simply: "It is our way."

Like the condors, the Chumash people are reclaiming their ancient ways—their culture could have been lost. "People are rediscovering our history," she tells me with excitement. A tribal museum—under construction as we speak—should be ready to open by the time you read this book. "In our upcoming museum, we will be able to tell our stories. We hope to bring our children there to show them our ancestors' way of life."

Those ancestors have, like AC-8, passed on to the spirit realm. But their message is even more urgent today, she says. They challenge all of us to ask important questions: How can we live to be better neighbors in this world?

The answers are all around us, says Jacy. Watching the zookeepers and researchers feed the baby condors in captivity, she realized, "If human kindness can continue, the condor can be stronger and resilient—like my people."

We pass by AC-8 again as Jacy and I say goodbye. I look into the taxidermied bird's red glass eyes one last time and ask Jacy what, as a tribal educator, she'd most like to share with you, the readers of this book.

"You can be the person in our community to make the difference," she insists as we part. "Everyone can be a steward for the land, for us, for our winged friends, for our children."

VULTURES IN CULTURES

THE GREAT NINETEENTH-CENTURY naturalist Charles Darwin, father of the theory of evolution, was not a fan of vultures. He thought all vultures were disgusting, their bald heads "formed to wallow in putridity." The biblical books of Leviticus and Deuteronomy claim vultures are unclean and that the children of Israel should loathe them as an abomination.

But in other times and other places around the world, many species of vultures have been celebrated and revered.

It's no wonder. Vultures are strong. Vultures are tender. Some even say that vultures are beautiful. And they possess an almost magical power: by eating carcasses to feed themselves and their young, they can turn death into life.

The Mapuche people of Chile call the Andean condor, the largest American vulture, the "king of birds." They say he exemplifies the virtues of wisdom, justice, goodness, and leadership. The Andean condor's close cousin, the California condor, is one of the most frequently mentioned animals in Western Native American mythology. The bird is sacred to many tribes. According to a Wiyot story, the condor was the original Noah. He and his

Andean condor.

Vultures are revered as exceptionally attentive parents. At left, a turkey vulture cuddles a week-old baby; at right, a black vulture with a slightly older chick.

sister were the only survivors of a great flood, and from their union sprang all the people today living on the earth. In California, several tribes believe the condor's feathers have healing powers; medicine men and women use the feathers in their ceremonies, and spiritual leaders evoke the condor's power in special dances.

In ancient Egypt, which was graced with at least five species of vultures, the birds' image adorns royal tombs and decorates temples, often holding divine seals in its feet. One species was singled out for special honor: The griffon vulture symbolized the goddess Nekhbet. In Egyptian art, Nekhbet was often shown holding her wings outstretched to shield a pharaoh, or floating above a young king. Her image is all over King Tut's tomb, from the walls to the coffin to his jewelry.

But in addition to protecting powerful kings, the griffon vulture also watched over

The king vulture ranges widely across Central and South America.

women in childbirth. The symbol for "mother" in Egyptian hieroglyphics is a vulture—because vultures are such good mothers. In fact, a fifth-century Egyptian priest, Horapollo, records the belief that all vultures were female. The wind, he believed, fertilized their eggs.

The lammergeier, also known as the sheep vulture or vulture eagle, was the sacred animal of the Scythians, an ancient tribe of nomadic warriors who lived in the western and central Eurasian steppe. The Germanic Goths, too, adopted this huge, powerful bird as their symbol. These vultures eat mostly bones, which they carry and drop in particular places to break them into pieces; they also seek out red soil, with which they dye their own feathers, giving them a fiery hue. (No one is sure why they do this; it may be for the same reason people use cosmetics.) Because of these traits, ancient, war-loving cultures may have thought the lammergeier was a fearsome predator.

But other peoples revered vultures not just for what they looked like, but for what they really are. The Cherokee peoples of the American southeast called the turkey vulture—the most common vulture in North America—the "peace eagle." They understood that, though its sharp beak and soaring flight resembles that of a bird of prey, the turkey vulture does no one any harm.

A number of human cultures have recognized vultures as important cleansers who consume death to produce life. The Latin name for the turkey vulture—*Cathartes aura*—recognizes this, too: *Carthartes* means "purifier" and *aura* means "golden."

Vultures, many cultures hold, honor those whose bodies they eat: by taking the flesh of the dead as their food, they carry the bodies of the departed with them on their wings, straight into heaven. In Tibet, for the past eleven thousand years and continuing to this day, many people offer the bodies of their dear departed to vultures in a practice known as sky burial. Human corpses are left on a mountainside to be consumed by vultures, the sky dancers.

Similar customs were practiced in Iran and, until recently, in India among the Zoroastrians. This practice seldom happens anymore in India because its vulture populations plunged in the mid-1990s. The vultures had been eating dead cows that had been treated for infections with the anti-inflammatory drug diclofenac. The country finally banned the drug in 2006.

The situation for vultures is even worse in parts of Africa. One of the continent's eleven vulture species was recently declared extinct, and seven others are endangered. Ironically, the population crash is largely because of yet another service vultures perform. When poachers murder elephants and rhinos, circling mobs of vultures often alert game wardens to the carnage, so the criminals kill these messengers. Poachers take the tusks and horns and poison the carcasses of elephants and rhinos they leave behind.

Happily, though, a movement is building to rekindle respect and reverence for vultures around the world. Ever since the first International Vulture Awareness Day in 2006, on the first Saturday in September, a growing number of zoos, animal parks, and conservation organizations publicize vultures' plight and powers with bike rides, talks, exhibit booths, and field expeditions. Check to see if your local zoo or wildlife center has a celebration you can attend!

A view from the nest.

Chapter Eight:

CHICK FLICK

THE FLUFFY GRAY CHICK with the pink, bulging crop looks as contented as a baby bird Buddha. But then something catches his attention. He begins eagerly waddling forward. He shivers his oversize downy wings. He throws himself forward—and lands in a face-plant in the fine sand of the floor of his cliffy nest cavity. The reason for his behavior is soon evident: one of his parents has arrived with another meal!

Estelle, Tia, and I are able to observe the chick as clearly as if we were actually inside the nest at Hutton's Bowl. But we're miles away—sitting in swivel chairs in the shady comfort of the camera room just outside the ranch house at Hopper Mountain National Park. "Being able to monitor nests remotely is so great!" says Estelle.

We can see the chick in such detail thanks to a video camera installed inside the nest. The images feed through a solar-powered wireless relay system. The system works like old-fashioned telephone lines, but the "wires" are invisible. The system sends data signals to radio repeaters mounted on poles, like cell-phone towers. A series of aligned dishes focus the signals. "With the cameras, we can see many more behaviors than we can see with our high-powered optics across the canyon," Estelle tells us.

This year, climbers placed cameras in three nests here at Hopper; Hutton's Bowl is the only nest that's still active. One egg didn't hatch; at the other, the chick, over thirty days old, was last seen stepping

Chicks' fluffy gray down provides camouflage inside the nest cavity.

Estelle watches the action
on the nest cam.

offscreen weeks before. It's presumed to have fallen off the cliff or to have been taken by a predator.

Thanks to collaboration between the US Fish and Wildlife Service, the Santa Barbara Zoo, the Cornell Laboratory of Ornithology, and the Western Foundation for Vertebrate Zoology, sixteen different nest cams have offered live-steaming video over the years at sometimes as many as three active nests at once. Another organization, the Ventana Wildlife Society, also operates a live-streaming condor cam.

"People across the country are keeping tabs on our nestlings," says Estelle. "It's pretty exciting!"

The parent bows to feed the chick, and when the big black bird turns, we can see its wing tag: it's a mom, Condor 289. It's a surprising scene—it looks like the big condor is trying to swallow the head of her offspring!

The parent, in fact, seems to be dragging the chick around by the neck. "But that's the baby not wanting to let go," explains Estelle. The chick wants more food.

And the noises! The cam relays audio as well as video, and this chick sounds like no baby bird you've ever heard before. Instead of cheeps and twitters, a baby condor makes a sound like a cross between a pig grunting and horse snorting—and it's *loud*. The mother responds in kind and then begins to gently preen her baby's wings. Then she looks up at something outside the nest. What does she see? A plane? An eagle looking to kill and eat the chick?

The parents' tenderness toward the baby begins with their attentiveness to the egg.

A parent nuzzles a chick.

Suddenly the scene goes black. Black feathers block the screen; dad has arrived. "If you were outside viewing through a scope," says Estelle, "you'd see him circle, land, fly below the nest, and then scramble up a bit. By using cameras alone, you lose that wider context. Are there red-tailed hawks flying around? Are the parents roosting in a tree nearby? There are always tradeoffs," Estelle observes—which is one reason, besides logistics and expense, why cameras won't be replacing direct observation anytime soon.

"But the cameras are great," Estelle continues. "We can record infrequent events like switchouts, when parents switch attendance

bouts. It enables us to keep tabs on chicks in a really efficient way." For instance, one camera recorded a chick with a leg injury. "We were able to send video clips to the vet and keepers at LA Zoo so they could see how the healing was progressing. Because they could see it was healing fine, we could leave the chick in its nest and monitor its medical condition without ever touching it. And our goal is as little disturbance as possible."

In some cases, nest cams reveal emergencies that demand hands-on action. Specially trained staff have climbed cliffs to evacuate chicks who've been poisoned by lead or stuffed full of microtrash. "The LA Zoo team can

remove the trash surgically and we can have the chick back in the nest within twenty-four hours," Estelle explains proudly.

When chicks are taken away for treatment, a human spends the night in the nest to prevent the parents from seeing that the chick is gone, which might cause them to abandon the nest. Petite Molly once spent the night in a cave so small her feet were hanging out!

Generally, parents wait nervously near the nest, flying from spire to spire. But occasionally the sight of a human entering the nest is too much to bear. "I once saw a mom condor strike a biologist hanging on a rope hundreds of feet in the air," Estelle tells us. "It's a tough

No, this chick is not dead! Vulture chicks tend to sleep on their stomachs.

just to carry those big wings around," she says. "It's so cute!"

But, of course, Estelle is here to do more than just admire the adorable chick. She's taking data, ticking off on her checklist all the signs that the chick is healthy:

Is the chick bright, alert, and responsive? Is the head healthy? Are the nostrils clear, beak intact, without wounds? Are the wings OK? No drooping? Are the legs sound? No dragging? Are the parents healthy? Does the chick have at least one positive interaction with a parent during the day?

Because all the video is archived, Estelle can also review past videos—and, conveniently, fast-forward through long stretches when the baby is alone and motionless. Of course, you can't do that when you're behind a scope under the hot sun!

We would have been delighted to get just a glimpse of the Hopper Mountain chick the day I spent with Erin and Kathy at the Snag Ridge observation point. But still, that day wasn't for naught, Estelle assures me. During those two hours, the parents did not come to the nest, and the chick did not step outside. That's important data.

"Technology isn't everything," she reminds us. Plus, the nest cams can't be installed everywhere. They are expensive to mount and

maintain; also, some nests are too remote to reach. And though the cameras function well, glitches still happen. The view from one nest cam was completely obliterated when a chick's poop covered the camera lens!

Every tool at hand is important to the condor recovery effort. For instance, says Estelle, "GPS is great because it tells you direction. There are even subtleties of behavior that on a GPS might suggest the bird is foraging. And as the condor population grows, without technology you would be looking at an exponentially larger staff size. Technology allows us to be

situation, being struck with that big beak. All you can do is wait it out till it stops." The biologist had a serious bruise from the attack, but no blood was spilled. "Generally, parents take it in stride. And when the chick is placed back in the nest, the parents immediately take care of it."

Onscreen, the dad and mom greet each other with soft snorts. The dad and chick twine necks; he rubs his baby's shoulder with his beak, then begins to preen the fluff. The baby plops down as if exhausted. He probably is, Estelle says.

"It takes a great deal of energy just to grow," she explains. The nestling shrugs to reset his wings. "It almost looks like it's tiring

A curious chick looks up at the nest camera.

in multiple places at once. But there's nothing like seeing the behavior yourself," she says.

No single method, no one technology, can do it all. The nest cams have proved an essential tool in helping condor staff guard the lives of these precious, vulnerable chicks. "Before we started nest guarding, the chief cause of death for chicks was eating microtrash," Estelle explains. "But we can *do* something about microtrash," Estelle adds. If parents bring trash to the nest and it's seen on the nest cam,

the condor team can even dispatch a climber to clean it up.

Right this minute, the Hutton's Bowl chick, in fact, is picking up some object—it looks like a stick—with its beak. But no trash is in sight. Just attentive, affectionate parents and a contented, well-fed, curious chick. Though Estelle knows that this chick—and all the California condors flying free—face many dangers ahead, at this moment, at this particular nest, all is right with the world.

Estelle finishes up her data sheet. "The parents look healthy; the chick looks healthy. The chick is on track. Things look really good!"

For now, that is. But what about tomorrow? Estelle has another place to take us before our time together is over. She plans to introduce us to more youngsters who live in condor country. And in them, we may glimpse the future.

This is the kind of nest-cam footage researchers always hope to see: a healthy, happy condor family.

Chapter Nine:

CONDOR KIDS

AT MOUNTAIN VISTA ELEMENTARY School in Fillmore, California, the third graders in teachers Kelly Myers's and Tanya Gonzales's classes are excited—and not just because it's the last day of school. They're getting a special visit today from two of their heroes: Estelle and Nadya.

They know the biologists because they've met them during their visits to the zoo and seen them on videos they've been watching. They and all the third graders throughout their school district have been studying a unit on condors for the past six weeks. "It's the pride and joy of the year for them," says Amber Henrey, the coordinator for the school district. Condor dioramas made in shoeboxes adorn one long table ("The condor's habitat is rocky forest," proclaims one label), and 8 x 10 billboards with slogans like SAVE THE CONDORS, NO LEAD, and PICK UP MICROTRASH/

DON'T DROP IT! hang on the wall behind it. Along other walls stretch two life-size outlines of the California condor—complete with many dozens of individual black wing feathers made of construction paper.

"We have the scientists presenting today!" Mrs. Gonzales tells the sixty students who have gathered in her classroom for the event. "It's Nadya and Estelle. Can you give them a hand?"

The kids all erupt in applause.

"Thank you," says Estelle. "I know you know so much about condors! You know more than my parents know about condors!" She invites them to show off their knowledge. "How many California condors are there in the world?"

Hands all over the room shoot up. Estelle calls on an eight-year-old girl with long black hair. "Four hundred and fifty," she answers.

"That's great!" says Estelle. "And that number is changing all the time. Actually there were 463 at the end of last year. And we have just under eighty living in your neighborhood of Southern California."

Next, the kids review some natural history: Estelle asks, "How do they have their babies?"

"EGGS!" the whole room cries at once.

"And how many eggs does a pair lay?"

"ONE!" the kids shout.

Estelle's impressed. The kids know how long condors nest (six months). They know the eggs have already hatched. They know that condors are feeding their baby chicks now—practically right around the corner. And the kids can answer perhaps the most important question about condors:

"How can we help them?" Estelle asks.

Again, hands go up all over the room.

Mountain Vista Elementary School sits below the Sespe Condor Sanctuary.

"Get rid of lead bullets!" says a boy wearing blue glasses.

"Clean up the microtrash!" says a girl with pigtails.

"So what is microtrash?" Estelle asks—and lots of kids answer at once:

"Gum wrappers."

"Shiny stuff."

"Glass."

"Bottle caps."

The kids understand that if you have a belly full of trash, you can't have a belly full of food—and you starve to death. The microtrash might not even get to the belly. It might just jam up in the bird's crop. Then the baby bird can't swallow. Any food it takes in just sits there and rots.

"And what can we do about it?"

"Clean up the microtrash!" the kids yell again. To them, the phrase has become like a refrain.

"Yes!" cries Estelle. "Every time you go to the beach or the park, pick up three pieces of trash—if everyone did that, it would make a big difference! And," she adds, "because you care about condors, now Nadya is going to give you the very latest news about them."

Nadya steps to the front of the classroom. "In 2015," Nadya says, "we had a record-breaking ten nests here in Southern California. But this spring we broke that record. We have *twelve* nests—more nests at any one site than we've seen since we reintroduced the California condor in the 1990s. This is really exciting!"

The kids agree. Soft *ohs* and *yays* sweep through the room.

"A lot of those nests are right in Fillmore," Nadya continues. "A few are way out—in Sequoia National Park up north; some are in the Santa Barbara backcountry. One is in Bitter Creek. This means the condors are moving. And we have a lot of new condor parents. It'll be fun to see how they do."

Those twelve nests, Nadya continues, have yielded eight chicks. Some of them are now more than sixty days old. "And it's at this point that the parents start bringing micro-

Estelle and Nadya speak with the class. On the board in front, the students' artwork shows the size difference between a golden eagle, a turkey vulture, and a condor.

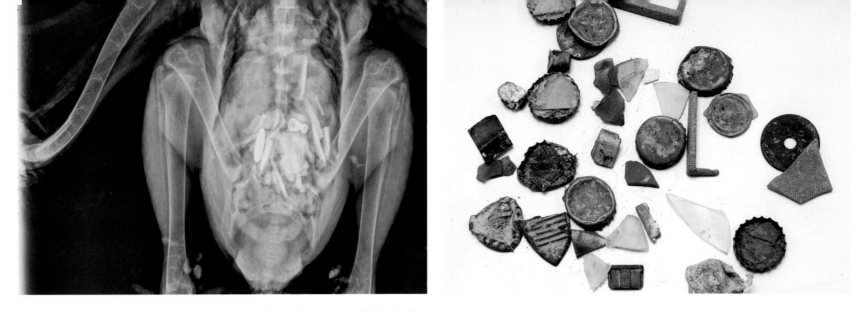

ABOVE LEFT: An x-ray of a chick treated at the LA Zoo shows the microtrash the chick ate.
ABOVE RIGHT: The Santa Barbara Museum holds collections of microtrash taken from the nests of condors 98 and 155.

trash into the nest." And that makes what the kids are about to do even more important.

Estelle steps in and makes the announcement. "I know we could talk all day," she says, "but we want to get outside and start picking up microtrash!"

The children instantly mobilize. They've done this before. They line up before Mrs. Gonzales so each child can take a plastic glove. "Don't pick up glass," she warns them, "but if you see glass, call us and we'll pick it up for you. And don't let the gloves blow away!"

With teachers carrying plastic basins for the refuse, the kids troop outside to the big schoolyard. It's near a local outdoor market and lots of trash blows in and gets caught in the wire fence along one side of the school. The kids have scarcely exited the door when a boy

Kailey Magana peers through the binoculars, looking for condors.

The students found and collected all this trash inside the fenced area of their schoolyard. Now a condor parent won't bring it to its baby in the nest.

wearing a sweatshirt with a star on it shoots out his arm, pointing. "Oh! I see microtrash!"

"Oh yeah, says his classmate, "I see some over there!"

The kids disperse over the schoolyard like an armada of vacuum cleaners, bending over and commenting on their finds:

"Oh gosh, I found a juice straw."

"I found a broken bottle cap—it's sharp!"

The same kinds of items have been found in condor nests and inside the bodies of condor chicks. The list goes on:

Flashlight batteries. Pop tops from metal cans. Screws and nails.

"We've seen condors eat and regurgitate screws and nails in nests," says Nadya. "We've seen them ingest wire. They eat pieces of plastic toys. At one nest I visited, I kept pulling out yellow plastic letters from some toy—at two different entries to the same nest. The condor parents kept coming back to wherever they had found that trash."

Knowing they can stop that from happening to a condor family has made a big impression on the kids at Mountain Vista. "Me and my brother, we don't leave trash around the yard," explains Jose Rodriguez. "We pick it up and throw it away!"

"It makes me happy to pick up the trash so it won't stick in their throat," says Julie Luna, who's almost nine. "Look at how many things I've found!" she says proudly, putting a plastic-gloved handful of bottle caps, cigarette butts, and plastic wrappers in one of the basins the teachers brought out for that purpose.

Eight-year-old Kailey Magana can't reach the ground from her wheelchair, but she's keeping an eye out for condors with her binoculars. She loves birds—she has a pet canary at home. So far she has seen a turkey vulture and a red-tailed hawk. But she really wants to see a

Students display a great haul of microtrash: five bins in just half an hour.

condor's nest," Estelle tells them. "Great job!"

She's deeply moved by what these sixty children accomplished. "This inspires me," she says. "Conservation can be a tough road, you know? But then you have a morning like this morning. These kids are healing the earth, cleaning up what other people left—and they're happy to do it for the condors."

Before we leave, Estelle, Nadya, Tia, and I take one last look at the paper condors taped up on the classroom wall. At close range, we see that the black feathers of their outstretched wings are fashioned from construction paper cutouts, tracings of the many small hands of the artists.

The fate of the California condor really is in their hands—and now it's in your hands, too.

condor. "They have big wings and I like them!" she says.

"Helping condors feels great," says nine-year-old Kason Hayes. He also loves the golden eagle and the turkey vulture. But when he grows up, he announces, "I want to be a research biologist for the California condor!"

At the end of just a half hour of collecting, the children have nearly filled five basins with trash. "So much of that could have been in a

TIMELINE

2.6 MILLION YEARS AGO–11,000 YEARS AGO: California condors thrive from Western Canada to northern Mexico, from Texas to New York to Florida. The birds feast on dead giants, like prehistoric ground sloths, camels, horses, and mountain goats.

AROUND 10,000 YEARS AGO: Condor range shrinks as humans arrive and start to kill off much of the available prey. The Pleistocene giant mammals go extinct in short order.

EARLY 1800s: Spanish settlers arrive, and their cattle and sheep provide a new bonanza for California condors. In the single year of 1834, in Baja California alone, Spanish colonists kill 100,000 domestic livestock for their hides and fat, leaving the rest for scavengers.

1840–1890: As more people flood into California, condors are killed for sport and science. During the Gold Rush, "Forty-Niners" maliciously shoot condors, often just for "fun." Even naturalists shoot condors for scientific specimens and raid their nests to sell their big eggs to museums and private collections.

1927: California condors are now so rare that when naturalist Robert Easton spotted two in the Santa Barbara backcountry, he wrote to the Smithsonian Institution in Washington, DC, to report his exciting find.

1937: An estimated 100 condors survive. Thanks to lobbying by Easton and the Audubon Society, a 1,200-acre sanctuary for condors is established in Santa Barbara.

1938–1948: The first formal scientific study of California condors is made. The Audubon and the University of California fund UC Berkeley graduate student Carl Koford's research. Condors are now rare outside of Santa Barbara and Ventura. Koford estimates the population to be about sixty. He suspects shooting by farmers, raiding of nests by museum collectors, and nest disturbance by photographers are to blame.

1949: After discovering that Andean condors lay a second egg if a first egg disappears, San Diego Zoo Safari Park seeks state permission to take condor chicks from two wild nests. The hope is that, while the two chicks grow up safely in captivity for later release in the wild, both sets of parents might lay two more eggs. But the permit is denied.

1963–1964: Audubon follows up on Koford's work by commissioning a study by two rancher friends of his. They estimate forty California condors remain—a decline of one-third since 1946. Cause of death is thought to be from shooting as well as from poisoning with strychnine and 1080, which are meant to kill predators and rodents.

1967: California condors are federally listed as endangered. The cause of their decline is still unknown. Could it be habitat destruction as wild lands are converted to housing developments? Maybe it's the pesticide DDT, which is known to thin the eggshells of nesting eagles? Something else?

1975: The California Condor Recovery Program is established by the US Fish and Wildlife Service. Lead poisoning from bullets in scavenged carcasses now emerges as the main suspect causing most birds' deaths.

1982: Fewer than twenty-five condors remain on planet Earth. Scientists begin capturing the remaining wild population in an effort to protect the birds and to begin captive breeding.

1987: The last wild condor, AC-9, is captured at Bitter Creek National Wildlife Refuge.

1988: San Diego Zoo Safari Park's breeding facility celebrates the first successful captive breeding since the condors' extinction in the wild.

1992: The first captive-born California condors are released into the wild at Sespe Condor Sanctuary.

1994: After complaints that ex-captive adolescents are haunting hikers and hang gliders and invading homes, all released birds are recaptured and returned to captivity. Zoos and sanctuaries revise their captive-rearing techniques.

1995: Releases begin again in Southern California.

1996–1997: Captive-bred birds are released at new sites in Arizona and in Big Sur, California.

1998: In captivity, condor chicks are now hatching at a rate of nearly twenty babies per year.

2001: For the first time since their wild extinction, condors attempt to nest in the wild.

2002: AC-9, the last condor to be captured from the wild, flies free again in Southern California. Condors are also released in Baja California, Mexico.

2003: The first condor chick successfully fledges in the wild in Arizona.

2004: The first chick fledges from a wild nest in California.

2008: More California condors fly free in the wild than live in captivity.

2013: Thirty leading scientists, doctors, and public health specialists from across the nation sign a consensus statement calling for an end to the use of lead ammunition everywhere—to protect *all* wildlife and people.

2019: California law takes effect, phasing out lead ammunition.

EPILOGUE: WHERE ARE THEY NOW?

Just before we went to press with this book, we checked up on some of our condor friends to see how they were doing. Here are Estelle and Nadya's reports:

Condor 20, the calm male who helped Joseph teach newbies on the condor team how to conduct health checks, fledged his first wild chick with his mate, 654.

526, the female who underwent a health check that same day (the volunteer who held her said she was "so pretty I can't handle it!"), was found shot near Bitter Creek National Wildlife Refuge in 2018, and later died. A reward of $15,000 was offered for information leading to the arrest of her murderer.

The Orchard Draw Chick, 895, is now a free-flying member of the Southern California flock. His parents nested again in the very same cavity in which he was raised, and their new chick had just fledged.

The Devils Gate Chick, 871, who escaped from the Thomas Fire in 2018, has a malfunctioning transmitter and hasn't been detected in a year.

771 and 480 (the condors the author held during their health checks) are both well. 480 did not nest during the 2019 season, but the condor team is hoping he might settle down with a mate and tend a nest of his own during the next nesting season.

Hopper Canyon dad 509 and mom 161, whose nest Sy, Erin, and volunteer Kathy watched in the spring, are still paired. Their chick is flying and well.

The Hutton's Bowl chick, who we watched on the chick cam, is doing well, and mom 289 and dad 374 are still paired.

And at Santa Barbara Zoo and at the LA Zoo, all the condors we met in these pages—Condor 174 and her young mentee, 603, ambassador Dolly, and surrogate mom Anyapa—are still alive and well. Come visit them if you stop by at either zoo!

WHAT YOU CAN DO

SUPPORT CONDOR PROJECTS. See the list on page 87 for organizations supporting the condor comeback. All need donations!

TELL OTHERS. Share your knowledge about condors with your family, your friends, your teachers. Can you write a report on condors to present before your class or school? Can you get your scout troop involved? Gather friends and neighbors to join your efforts!

CLEAN UP MICROTRASH. In condor country, bits of human garbage, especially plastic and metal, can kill a condor. But no matter where you live, picking up trash may save a life—whether it's a sea turtle that won't choke on a floating plastic bag, thinking it's a jellyfish, or a skunk that won't smother from getting her head stuck in a small-mouthed, wide-bodied Yoplait yogurt container.

REDUCE PLASTICS. Carefully choose what you buy so you don't add to the mountain of plastic trash. Recycling plastic is becoming increasingly diff-icult. So, instead of using plastic bags, carry purchases home from stores in a nice cloth sack (sometimes even buying a cloth bag will benefit wildlife!). Shun balloons as party decorations. Choose drinks packaged in boxes, glass, or cans. Skip the plastic straw. You can think up even more ideas—and share them with your family, friends, school, and local newspaper. Remember, when you or your family pay for a plastic item, you're rewarding its manufacturer for making something that kills wildlife and won't degrade for thousands of years!

WRITE TO YOUR LEADERS and ask them to support a phase-out of lead ammunition and fishing tackle in favor of less harmful alternatives. Tell them how serious the problem really is, and how much protecting animals matters to you. Send a copy to your local newspaper. Don't be afraid to include your age, to teach leaders that both condors *and* kids matter!

BIBLIOGRAPHY

BOOKS:

Caras, Roger. *Source of the Thunder.* New York, NY: Little, Brown, 1970.

Fallon, Katie. *Vulture: The Private Life of an Unloved Bird.* Lebanon, NH: University Press of New England, 2017.

Koford, Carl. *The California Condor.* New York, NY: Dover Publications, 1966.

Mee, Allan, and Linnea S. Hall, editors. *California Condors in the 21st Century.* Fayetteville, AR: Nuttall Ornithological Club and the American Ornithologists' Union, 2007.

Mesa, Robert. *Condor: Spirit of the Canyon.* Grand Canyon, AZ: Grand Canyon Association, 2007.

Moir, John. *Return of the Condor: The Race to Save Our Largest Bird from Extinction.* Guilford, CT: Lyons Press, 2006.

Snyder, Noel, and Helen Synder. *The California Condor.* London, UK: Academic Press, 2000.

Wilbur, Sanford R. *Condor Tales: What I Learned in Twelve Years with the Big Birds.* Gresham, OR: Symbios, 2004.

FILM:

Scavenger Hunt. Director: Eddie Chung. Producer: Matthew Podolsky. Cinema Libre Studio, 2013.

TO LEARN MORE

WEBSITES AND WEBCAMS:

Santa Barbara Zoo: www.sbzoo.org/conservation/condor-cam

US Fish and Wildlife: www.fws.gov/sno/es/CalCondor/Condor.cfm

WILD CONDOR CAMS:

Southern California, www.cams.org/chanel/49/California_condor

Central California, www.venturaws.org/condor_cam.html

Pacific Southwest, www.fws.gov/cno/es/CalCondor/CondorCam.html

ZOO CAMS:

San Diego Zoo Safari Park, www.sdzsafaripark.org/condor-cam

Oakland Zoo, www.oaklandzoo.org/webcams#condor-cams

CONDOR CURRICULUM DEVELOPED FOR SOUTHERN CALIFORNIA SCHOOLS:

www.condorkids.org

Help celebrate International Vulture Day: www.vultureday.org

ORGANIZATIONS RELEASING AND MANAGING CONDORS IN THE WILD. THEY NEED YOUR SUPPORT!

- Ventana Wildlife Society, www.ventanaws.org
- Peregrine Fund, www.peregrinefund.org
- Pinnacles National Park, http://pinnacles.org. (To donate go to Friends of Pinnacles, http://pinnacles.org/about/donations.html.)
- USFWS Hopper Mountain NWR Complex, www.fws.gov/refuge/hopper_mountain (click on Get Involved); Friends of California Condors Wild and Free (www.friendsofcondors.org); the Great Basin Institute (www.thegreatbasininstitute.org) helps fund internships; and Santa Barbara Zoo's condor field program (www.sbcondors.com/recovery/sb-zoo).
- In Mexico: ENDESU, www.endesu.org.mx

ZOOS AND OTHER FACILITIES THAT HOUSE, MEDICALLY TREAT, OBSERVE, AND/OR BREED CONDORS:

- Chapultepec Zoo, http://data.sedema.cdmx.gob.mx/zoo_chapultepec
- Cornell University, www.org/guide/California_Condor/overview
- Friends of California Condors Wild and Free, www.friendsofcondors.org
- Los Angeles Zoo & Botanical Gardens, www.lazoo.org
- Oakland Zoo, www.oaklandzoo.org
- Oregon Zoo, www.oregonzoo.org
- The Peregrine Fund, www.peregrinefund.org
- San Diego Zoo Global Wildlife Conservancy, www.endextinction.org
- Santa Barbara Zoo, www.sbzoo.org
- Phoenix Zoo, www.phoenixzoo.org
- California Living Museum, http://calmzoo.org

ORGANIZATIONS SUPPORTING NEST CAMERAS:

- Cornell Lab of Ornithology, www.birds.cornell.edu/home
- Explore, https://explore.org

ORGANIZATIONS SUPPORTING NON-LEAD ALTERNATIVES.

- Institute for Wildlife Studies, www.iws.org
- North American Non-lead Partnership, http://nonleadpartnership.org

ALSO SUPPORTING THE CONDOR PROJECT:

- Santa Barbara Museum of Natural History, www.sbnature.org
- Grand Canyon National Park, www.nps.gov/grca
- Redwood National Park, www.nps.gov/redw
- Sequoia & Kings Canyon National Park, www.nps.gov/seki
- Institute for Wildlife Studies, www.iws.org
- LightHawk, www.lighthawk.org
- Santa Ynez Band of Chumash Indians, www.santaynezchumash.org
- Smithsonian Institution, www.si.edu
- Tejon Ranch Conservancy, http://tejonconservancy.org
- The Wildlands Conservancy's Wind Wolves Nature Preserve, www.wildlandsconservancy.org/preserve_windwolves.html
- The Yurok Tribe, http://yuroktribe.org

INDEX

SCIENTISTS IN THE FIELD

Where Science Meets Adventure

Check out these titles to meet more scientists who are out in the field—and contributing every day to our knowledge of the world around us.

Looking for even more adventure? Craving updates on the work of your favorite scientists, as well as in-depth video footage, audio, photography, and more? Then visit the new Scientists in the Field website!

sciencemeetsadventure.com